Best Practices for High Impact Threats to Critical Infrastructure

Best Practices for High Impact Threats to Critical Infrastructure

Conference Proceedings of the
InfraGard National EMP SIG Sessions
at the
Dupont Summit
Auditorium, Whittemore House, 1526 New Hampshire Ave NW
Washington, DC 20036
December 2, 2016

Video-recorded sessions at
http://www.ipsonet.org/conferences/the-dupont-summit/dupont-summit-infragard-2016-videos

Edited by Charles L. Manto and Stephanie A. Lokmer

WESTPHALIA PRESS
An Imprint of Policy Studies Organization

Best Practices for High Impact Threats to Critical Infrastructure
Conference Proceedings
InfraGard National EMP SIG ™ Sessions at the 2016 Dupont Summit
December 2, 2016
At the Whittemore House
1526 New Hampshire, NW, Washington, DC 20005

Hosted and Published by the Policy Studies Organization
Edited by Charles L. Manto and Stephanie A. Lokmer
October 2017

Westphalia Press
An imprint of Policy Studies Organization
1527 New Hampshire Ave., NW
Washington, D.C. 20036
info@ipsonet.org

ISBN-13: 978-1-63391-611-1
ISBN-10: 1-63391-611-1

Cover design by Jeffrey Barnes:
jbarnes.design

Daniel Gutierrez-Sandoval, Executive Director
PSO and Westphalia Press

Updated material and comments on this edition
can be found at the Westphalia Press website:
www.westphaliapress.org

EMP SIG
Education. Communication. Power. Resilience.

TABLE OF CONTENTS

Preface and Acknowledgements

These 2016 proceedings represent the fifth annual conference of the InfraGard National Electromagnetic Pulse Special Interest Group (EMP SIG) on the 20th Anniversary of InfraGard at the Dupont Summit hosted by the Policy Studies Organization. The video recordings are available at: http://www.ipsonet.org/conferences/the-dupont-summit/dupont-summit-infragard-2016-videos

These particular conference proceedings are notable for the announcement by Congressman Trent Franks' House passage earlier on the same day, December 2, of the Critical Infrastructure Protection Act as Section 1913 of the National Defense Authorization Act subsequently signed into law on December 23, 2016. This law directs the Department of Homeland Security (DHS) to develop plans and outreach to prepare for high-impact threats such as EMP and extreme space weather.

The conference also reviewed nationwide activities of the EMP SIG, including a table-top exercise in Texas, the space weather conference workshop in Colorado, the Energy Tech conference in Cleveland, Ohio and EMP SIG publications. Most noteworthy is the release of *Powering Through: From Fragile Infrastructures to Community Resilience,* a planning guide for communities, companies, and government agencies to prepare for and mitigate widespread prolonged infrastructure collapse. Copies were presented to leadership within Congress, the National Guard and DHS at the conference. After the conference, the publication team began a process of expanding into nine additional teams to begin work on the more expansive second version, 2.0.

The conference segments also reflect the work of EMP SIG members in related efforts: the INCOSE Critical Infrastructure Protection and Recovery Working Group, the National Guard Bureau, and the Society for Disaster Medicine and Public Health (SDMPH). Key articles related to conference topics and a number of papers authored by EMP SIG members and conference presenters are reprinted from the January 2017 INCOSE Journal *Insight.* Given the iconic impact of these prolonged widespread disasters to health and medical communities, the prepublication copy of a *Resilient Hospital Handbook* by Dr. Earl Motzer, Dr. James Terbush and Mr. Charles Manto is also included. The Handbook, outlining approaches to resilient microgrids, is a work product of the NIPP Resilience and Security Challenge of the National Institute for Hometown Security. This work was also presented to the annual meetings of the SDMPH and the Milwaukee Regional Medical Center.

The EMP SIG wishes to thank the Policy Studies Organization (PSO) for its generous support of the conference and the assistance of many who helped with editing. We also wish to thank the National Guard Association of the United States who provided use of their facilities for workshops and tabletop exercises. Of course, the EMP SIG, as a nationwide Special Interest

Group of InfraGard, appreciates the strong support of the National Board and staff, local chapters across the country, its members across all the 50 states and territories, and the Federal Bureau of Investigation, who provides significant support to its InfraGard program. Together, all of these organizations have provided trustworthy venues and collaborations to foster information sharing leading to greater resilience in local communities nationwide.

I also wish to thank Mary Lasky, Cathy Feinman, Michael deLamare, Stephanie Lokmer for assistance in editing and document acquisition. Of course, the Policy Studies Organization team has been essential in the overall publication process and is due great appreciation from the InfraGard EMP SIG community.

As the outgoing Chairman of the InfraGard EMP SIG and its conference session organizer, I welcome you to join the new EMP SIG chair, Mary Lasky, and the two vice-chairs, Bill Harris and Steve Volandt in greater success for the coming year. For information on related topics, see documents on EMPcenter.org and to join InfraGard, see www.infragard.org.

Charles Leo Manto
EMP SIG Chairman Emeritus, InfraGard National
cmanto@stop-EMP.com

InfraGard National EMP SIG Guidance Document

The purpose

The purpose of the InfraGard National EMP (electromagnetic pulse) SIG (special interest group) is to foster communications and coordination that will address and mitigate the threat of a simultaneous nationwide collapse of infrastructure from any hazard such as manmade or natural EMP.

Method Focusing on Local Sustainability

The National EMP SIG will mobilize subject matter experts at the national level so that local InfraGard chapters can make use of them to help local communities become more sustainable in light of these threats.

Resources

Expert National Advisory Panels

The National EMP SIG will establish panels of leading advisors. Examples may include but are not limited to:

1. a **civilian-military liaison panel** enabling local communities to become more resilient so they can better support their local military and National Guard resources;
2. a **legislative and policy liaison panel** that can facilitate discussions between interested leaders at the national and local level with those in the private sector to identify and fill policy gaps;
3. an **education panel** that can facilitate research, development and education/training into the development of human resources needed;
4. an **investment panel** that might identify the capital support needed by local communities to enhance its sustainability;
5. a **media, communications and outreach panel** that can facilitate communications among EMP SIG members and between other stakeholders outside the SIG;
6. an **internal coordination panel** that would coordinate activities between the EMP SIG and other SIGs and committees within InfraGard; and,
7. a **communications liaison panel** that would share information about emerging communications technology and make use of it to further the activities of the EMP SIG and InfraGard.

Qualifications and Expectations of National Panel Members

Qualifications: Panel members will be chosen based on their leadership within their respective fields by virtue of knowledge, experience, roles, capabilities or relationships.

Expectations: Panel members will agree to attend an in-person meeting once per year, several conference calls during the course of the year and occasional email or phone correspondence with EMP

SIG leadership. However, given the volunteer nature of these roles, it is expected that their contributions while meaningful will be limited and offered on a best-efforts basis.

Appointment of National Advisory Panel Members: Membership to the national panels will be by appointment by the EMP SIG Chair (manager). The initial chairman of each panel, if needed, will be appointed by the EMP SIG manager and subsequently by vote of the panel members at intervals of their selection.

The InfraGard National EMP SIG leadership team includes:

1) The InfraGard National EMP SIG Chair (manager) who serves at the pleasure of the InfraGard National Members Alliance (INMA) Board of Directors.

2) The InfraGard EMP Guidance Committee composed of the INMA Chairman, the INMA President, the INMA Managing Director and an FBI HQ Supervisory Special Agent from the National Industry Partnership Unit (NIPU) who functions as an official liaison for the FBI to InfraGard on EMP matters.

3) The InfraGard EMP SIG Senior Management Team who is appointed by the EMP SIG Chair to serve the EMP SIG membership through the following positions:

a) The Vice-chair who is familiar with key EMP SIG activities and can take the place of the Chair during periods of the Chair's unavailability;

b) The Administrative Officer, whose role is that of records keeper and other administrative functions of the Senior Management Team;

c) The Strategy Officer, who will help develop and align EMP SIG policy with its activities;

d) The Finance Officer, who will help track financial activities to support the EMP SIG and its coordination with the INMA treasurer;

e) The Regional Outreach Facilitator, who will assist the Senior Management Team and its activities with local InfraGard chapters and Regional Reps;

f) The Liaison Panels Facilitator, who will support the work of the Liaison Panels and their interactions with the Senior Management Team;

g) The Subject Matter Expert Panels Facilitator, who will support the work of the SME Panels and their interaction with the Senior Management Team; and,

h) Other ad hoc committees that the Senior Management Team may deem useful from time to time.

EMP SIG Membership

Any InfraGard member in good standing may join the EMP SIG at the national level by indicating interest in participating in activities, mailings or communications designed for membership participation.

Members may either be designated as "working members" or "observers." EMP SIG members are encouraged to ask EMP SIG national leadership for help with their local EMP SIG activities.

Secure Communications between SIG Members

The InfraGard secure portal will be the primary means of communications between SIG members. EMP SIG leadership will also assist in providing other resources, including those that may be less secure, to supplement the official InfraGard communications and resources as needed on a best-efforts basis. This will include library resources and links to resources deemed to be of special value by the membership of the EMP SIG.

Other resources

The EMP SIG will be responsible for recruiting and raising resources necessary to perform its tasks subject to the normal and customary procedures and governance of the InfraGard National Members Alliance and InfraGard Members Alliances.

Governance and Activities of the EMP SIG: All activities of the EMP SIG will comply with the governance and ethics as required by the INMA Bylaws and any guidance provided by the INMA Board of Directors.

(Background: The concepts of this guidance document have been proposed by the founding SIG Chair (manager) and approved by the INMA and the FBI NIPU. See the initial authorizing letter for the EMP SIG and the initial EMP Committee/SIG proposal for additional background.)

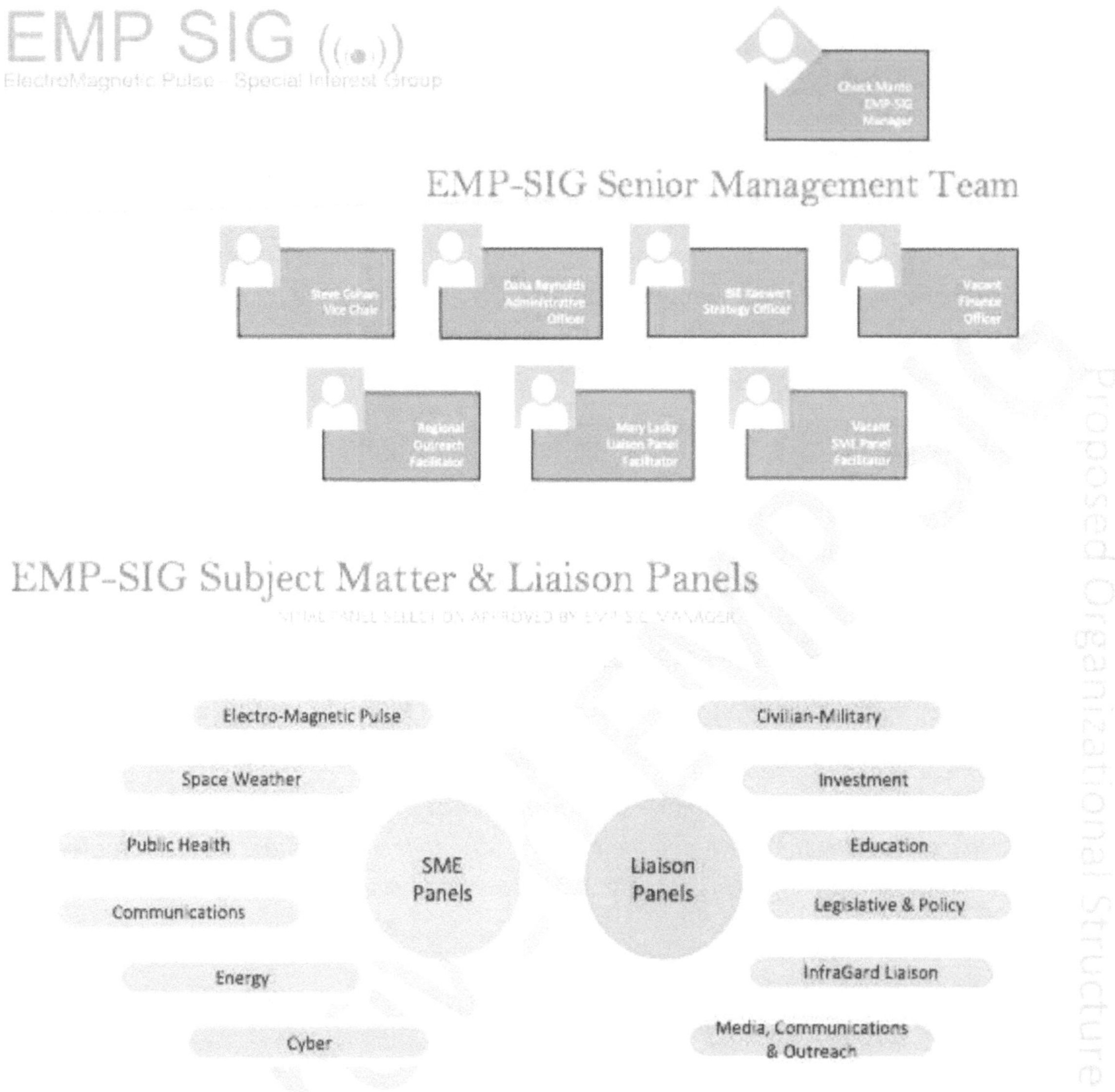
EMP SIG
ElectroMagnetic Pulse - Special Interest Group
Chuck Manto
EMP-SIG
Manager
EMP-SIG Senior Management Team
Vice Chair
Dana Reynolds
Administrative
Officer
Strategy Officer
Vacant
Finance
Officer
Regional
Outreach
Facilitator
Mary Lasky
Liaison Panel
Facilitator
Vacant
SME Panel
Facilitator
EMP-SIG Subject Matter & Liaison Panels
Electro-Magnetic Pulse
Space Weather
Public Health
Communications
Energy
Cyber
SME Panels
Liaison Panels
Civilian-Military
Investment
Education
Legislative & Policy
InfraGard Liaison
Media, Communications & Outreach
Proposed Organizational Structure

Conference Proceedings of the InfraGard National EMP SIG Sessions

Note: The time stamp notations found throughout the texts are indications of inaudible portions of the video presentations. For your reference, the corresponding videos will be posted online at: http://www.ipsonet.org/conferences/the-dupont-summit/ dupont-summit-infragard-2016-videos

Best Practices for High Impact Threats to Critical Infrastructure

Program Details

InfraGard National EMP SIG Sessions of the Dupont Summit

December 2, 2016

Auditorium, Whittemore House, 1526 New Hampshire Ave NW, Washington, DC 20036

The focus of this year's summit will be to examine current and emerging best practices in resilience needed to protect ourselves from high-impact threats to critical infrastructure.

(For secure DHS WEBCAST access on the day of event, go to the following web address and *register as a Guest*, preferably 15 minutes ahead of time: https://share.dhs.gov/empsig) HSIN requires Flash Player to be installed. Video or audio recordings of these sessions are expected to be posted by the Policy Studies Organization under the conference section of their website at ipsonet.org and affiliated YouTube locations. Empcenter.org may also carry portions on their website.)

Program Details

8:00 **Registration**

8:30 **Welcome and Introductions**
Mr. Charles Manto, INMA Board
Mr. Brad Brekke, Director, Office of Private Sector, FBI Director's Office

9:00 **Congressional perspective**
Hon. Roscoe G. Bartlett, PhD; Congressman Trent Franks

9:30 **PSA Program and ties to EMP SIG**
PSA Director Scott Breor

9:50 **Recent Developments in DHS and Space Weather Executive Order**
Hon. Caitlin A. Durkovich, DHS Assistant Secretary

10:20 **Panel: Recent State Legislation and Actions**
Mr. Ron Jacobus representing Sen. Bob Hall, Texas
Hon. Andrea Boland, Maine; Virginia, North Carolina

11:00 **Report from the New EMP Commission**
Dr. Peter Vincent Pry
Dr. George Baker

11:30	**"Powering Through" and the Dec 1 Workshop** Ms. Mary Lasky
11:50	**Implications for Best Practices** Mr. Charles Manto
12:00	**Lunch**
1:00	**Best Practice Needs for Disaster Medicine and Public Health** Dr. James J. James, Executive Director, SDMMPH
1:20	**Medical Panel – Best Practice Assessment Underway by DHS, NIHS Hospitals for Resilient Hospital Microgrid** Dr. Terry Donat, MD; Dr. Earl Motzer, Dr. James J. James
2:00	**Preparation for Best Practices for EMP-Protected Microgrids** Amb. R. James Woolsey; DHS SHARES Mr. Robert Lachenmayer, COO, Pos-EN Mr. David Geary, Power Analytics Mr. Douglas Kiesewetter, Carpe Solar Mr. William Kaewert, SENS
2:30	**Space Weather Executive Order** Mr. Robert Rutledge, Space Weather Prediction Center, NOAA
3:00	**Recent Power Grid Studies** Mr. Thomas Popik Mr. Bill Harris
3:30	**Regional Activities of INCOSE and EMP SIG** Mr. John Juhasz Comm. Peg Littleton Mr. Steven Volandt
4:00	**Power and Water Utility interdependencies Best Practices** Dr. Paul Stockton
4:30	**Cameo Presentations** Amb. Hank Cooper Mr. Jack Pressman
5:00	**EMP SIG Next steps**
5:30	**Closing**

Welcome and Introductions

Mr. Charles (Chuck) Manto introduces the fifth annual EMP SIG Dupont Summit Sessions at the 20th anniversary of InfraGard. He also introduces Dr. Earl Motzer of the INMA Board who will introduce Mr. Brad Brekke, Director, Office of Private Sector, FBI Director's Office

CHUCK MANTO: This morning before we begin, I'd like to first introduce myself and some of the people involved today. First of all, my name's Chuck Manto. I'm the Chairman of the InfraGard National Electromagnetic Pulse Special Interest Group, or what we call the EMP SIG.

For those of you who don't know InfraGard, it's affiliated and sponsored by the FBI, and our host from the FBI will speak to that in just a moment. But I certainly want to spend a moment to thank the FBI and InfraGard for all the support that they give us throughout the year and since we were formed five years ago in 2011. This is the 20th anniversary of InfraGard, and the fifth anniversary of the EMP SIG. Let's give a round of applause to the FBI. [applause]

I also have to mention that the Policy Studies Organization, we may hear from Dr. Rich later, is a gracious host for this every year. They subsidize and provide for us this great venue. They're responsible for the food that you're eating today. They help us with our publications, including the conference proceedings, which you received when you came in today. So let's also give appreciation to the Policy Studies Organization. [applause]

In addition, from time to time we have corporate sponsors. Today, just wanted to mention we have some great help from two companies – PosEN, which many of you may not know, but they're a microgrid company; and a data center company that does EMP protection, Cyber Innovation Labs. Thanks to them for providing meals yesterday and last night. [applause]

I also wanted to take a moment to recognize the National Guard Association of the US. They provided their facility last year, and this year again. It's been the third year in the row that we've been able to use their facility, and we have a great partnership with them. And I wanted to thank the National Guard Association of the US as well.

We have many special guests with us today we can't even enumerate. Just a show of hands – how many military or former military do we have in the room? Wow, look at that! [applause] Also, elected or former elected officials, raise hands. [applause]

We have with us today folks from the state of Maryland, which represents the Maryland Department of General Services and the Secretary today with the Assistant Secretary Leigh Williams and crew. She's here; raise hands. [applause] And Maryland companies have been at the forefront of trying to do something to resolve this today.

Throughout the day, you're going to hear– one of the purposes of this meeting that we do every year is the following: You're going to see some very fast-paced presentations. It's almost like going to Congress; we have five minutes to tell Congress what to do to fix the world. And you only get five minutes. You may have hours of information you need to share and a million pages of data, and you hand that to the staff. And that's sort of what we do today, very fast-paced presentations. Many of them are like TED Talks – boom, boom, boom, boom.

The reason for that is really simple. We try to bring leaders of the technology arena, from the private sector and the public sector, who are identifying new technologies and emerging threats that come from that technology, and then identify what we see as maybe emerging policy gaps. Because when these new problems arise, very often nobody's thought of it before, nobody's really in charge of that sort of thing. So here, we try to bring policymakers, technology leaders, managers of people who run organizations in the same room together because very often our busy schedules don't allow that to happen.

So today, I hope that in addition to getting a taste of all these different viewpoints today in a very superficial way, because we don't have enough time to do more than scratch the surface, we hope you appreciate each other. Take the moments that we have during lunch and other opportunities to introduce yourselves to each other and begin to either forge a new relationship or build on the ones that you already have.

So as we begin to do that this morning, what I would like to do is give the special appreciation to the FBI. And I'm going to have one of the InfraGard national board members, Dr. Earl Motzer, introduce them. I'm very excited about our presentation this morning because this is not a typical person from the Bureau. This is someone from the private sector who was recruited to the Bureau. So we're in for a special treat, just to get that perspective.

But this morning, which I'd like to do is introduce you to Dr. Earl Motzer. Dr. Motzer's been a CEO of a hospital for close to 50 years or so. I think we went to high school with Abraham Lincoln together. [laughter] But we have people in our group who have a wealth of experience. He's one of them. He's been involved in education programs. I'd like him to speak just in a sentence at least and tell us what your whole role is with HHS and what that all means, because many here may not understand that.

And then what I'd like to do is have Dr. Earl Motzer, who's on the board of the InfraGard National Members Alliance, to whom we at EMP SIG report. We are formed as a committee of the national organization, and we serve under the direction of their board of directors.

And so, what I'd like to do now is turn it over to Dr. Motzer, who will introduce our first speaker. [applause]

Mr. Brad Brekke

Director, Office of Private Sector, FBI Director's Office,

just having been introduced by Dr. Earl Motzer of the INMA board

BRAD BREKKE: Good morning.

ALL: Good morning.

BRAD BREKKE: Welcome on behalf of the FBI and the Office of Private Sector. I'll give you a little background on myself before I go into my remarks. As mentioned, I come out of the private sector as chief security office for Target Corporation for over a dozen years. Prior to that, I actually was in the FBI as a brick agent, actually doing real work, not headquarters work. And after leaving the Bureau, I practiced law for a short period of time, but quickly found out that did not make my heart beat faster, so moved on to corporate.

A couple years ago, Director Comey looked at their efforts around a private sector and determined it was time to formalize their efforts – and I'll speak a little bit to that in my remarks – and set up an Office of Private Sector with the mission of building a strategy and formalizing outreach to private sector engaging with the FBI. So that's why I'm here today.

I want to recognize the leadership of InfraGard, the leadership of the EMP SIG. I do appreciate what you do, and understand many of you have day jobs or other roles. So it's with great appreciation to see you in the room today for this event.

I want to tell the story of where the Bureau has been and where it's going through InfraGard because, as just mentioned, it is its 20-year anniversary, and I think that is significant on a number of accounts and use that story to tell the Bureau's story of today.

Let's go back 20 years. 1996. Let's think state of the world back then. Culturally, we had movies such as *Independence Day*, the first one, coming out; *Mission Impossible*, the first one. Music was big on Madonna and Sheryl Crow. The Brits had a set of issues overseas; they had mad cow disease and Prince Charles and Diana were in the midst of a divorce. In the US here, we were hosting the Olympics in Atlanta that summer. And our economy was actually growing remarkably back in 1996. The stock market had hit a high of 6000 with maybe one bad news story – Apple stock was at a 10-year low, under $18 a share.

The tech world was starting to take off. It was leading up, of course, to the bubble we know about now, but it was on fire also. Entertainment was big. DVDs had just come out. And again, another first, hottest game going, Pokémon. Computers were becoming mainstream; Microsoft introduced Windows NT 4.0. Computers were also becoming powerful. IBM's Deep Blue defeated chess champion Garry Kasparov in a chess match; first time ever a computer beating a human at that level. And on a practical day-to-day matter, email replaced postal service for the most messages being transmitted on a regular basis.

For the FBI, a few things were happening that I think are very significant. One, as even now with the terrorist acts, we had the bombing at the Olympics that summer. We were able to actually arrest the Unabomber Ted Kaczynski that summer. Then a very [4:22] an event of great significance as we now know, Osama bin Laden was expelled from Sudan and went to Afghanistan that year.

In Cleveland, a group of FBI agents – some of you know Van Horn and other people going back – got together in 1996 with the private sector, with academics and with business leaders, and began to look at how to deal with the emerging, and I'll call it a cyberthreat, keeping in mind how primitive the threats were back then; they were primarily intrusion or breaches. And not only what they were, but how to counter it, how to prevent it, how to mitigate them.

From the FBI, this is rather unique from a couple perspectives. One, cyber was brand new. It was a threat that really wasn't mapped well, it wasn't high on what would later be our threat band. And the second thing, this was a public/private partnership. I served previous in the Bureau. At that time, the only time you dealt with the private sector was to interview them for a 302 or to serve them with legal process. You did not work with them.

So this is rather unique, how it started InfraGard in 1996. I would describe it as over the horizon. It was literally ahead of the threat because cyberthreat had not emerged to what it's become today.

Now, you also have to keep in mind, if you go back at that point, just to give you context, at that point in time there was no US-CERT, there was no DHS or NCCIC, there were no ISACs, no service providers like FireEye and CrowdStrike. There was no focus on cyber. And this group of public and private began that back then.

Let's fast forward to today. I think we would all agree the world has changed much. The threats have evolved immensely, and the FBI is evolving with them. Director Comey and the FBI leadership defined our mission to protect the American people and uphold the Constitution; sort of classic for us. But they also defined our vision, what we aspire to. Let me read it to you: Staying ahead of the threat through leadership, agility and integration. Let me highlight a couple things about this vision:

One, ahead of the threat. Something we learned post-9/11 to be with counterterrorism, but now we're applying across the Bureau as a whole – the mindset to be out in front.

Through leadership. That's internal to us, to find the right leaders that take us there.

Agility. This one is challenging, and I had some conversations before I came up here. Having government pivot quickly, like the private sector can, is a challenge. But it is part of the intent.

And integration. What is meant by that, by the Director, is integrating within the FBI, within USG, with our partners, with the public sector, law enforcement, which we did after 9/11 and developed all sorts of fusion centers, JTTFs. And now the Office of Private Sector with the private sector to integrate so that we [08:29] in a formal, strategic manner. It is very similar to the way InfraGard started when you think about it. InfraGard was the nucleus of that idea – public/private partnerships ahead of the threat.

So what we are looking at across InfraGard and other programs, legacy programs, how do we reinvent ourselves. We're working with InfraGard leadership to not only reinvent, but perhaps reimagine what these relationships, what these partnerships can yield.

We're also looking at new and different ways to interact. The world's changed. Many of these companies are different than they were before. It is easy to look at a Boeing and understand how to work together with them; it can be challenging to look at a Google and understand how to work with them. So we're exploring different ways to work across the industry.

The reason is simple, I would say. And having been in the private sector, I understand it – you possess tremendous expertise and capability and experience. The private sector is key to many things. In fact, I think the deputy director said it well the other day. He was addressing the US Chamber's national security group. And these were not part of his planned remarks, they were part of a Q&A session. He was asked, What role does private sector play with cyberthreat? And there were different nuances to the question – should it be more, should it be less? He didn't really answer the question asked. What he said was this: From the FBI's perspective, from his view, what the private sector has is innovation, the ability to see problems, to move quickly, to fashion solutions from that. What the FBI has is information around the threats and others. What we need to do is come together, not just to share information in a classic sense, but collaborate. What we need is collaboration, which by definition is coming together to solve a problem.

And that is the mindset of the FBI, of the Office of Private Sector, of what we're seeking to do now with InfraGard and other programs, is how to better collaborate, how to come together.

I maybe want to put this in context as to why this is so important. I'm going to use a study done by an author by the name of Joel Kotkin. Joel looked at cities and he asked himself a question – he's also a social historian and futurist – why is it some cities thrive and are vibrant and last through centuries and other cities don't; they come and then they fade.

So he studied thousands of years of ancient history, and the book's short, and actually the answer is in the first chapter, which is probably why I read it. [laughter] After studying he came up with three attributes that he identified that define a vibrant, successful, long-lasting city. And I would offer that these same three apply not just to cities, but to our nation. He found this about them:

That they had something sacred. They had either a temple, a river or a seat of government, as he defined it, sacred in their center. I think of no better example than this city right here. That was the anchor.

The other two things he found repeatedly through all of history was they were safe and they were busy. It was a nexus with security and commerce that allowed them to thrive and be vibrant. And I would argue that that holds true for our country. We are recognized as the number one world power, still, although many are trying to overtake us. And I would say there's two reasons for that.

One is our US military; it is above all others. The second is our economy; it is an economic engine that all seek to obtain. There's a reason why the dollar is a global, international currency. I would also say there's many threats against that economy from a counterintelligence, from a cyber, from even a terrorism side.

Back to Joel's study. We need security, whether that's federal level or local level, working with you, the private sector, commerce, to ensure that that is not jeopardizing to our country. That is what drives us in our effort to reach out to the private sector and explore the ways we can work closer as we look to the future and whatever threats or changes that entails.

I would also like to highlight this group itself. It's an excellent example of that model. When the private sector has come together with the government, across all different entities, we work to protect our country. And I'll point out, in 1996, that mission was not envisioned by the InfraGard group, but grew out of it through the efforts of the people in this room.

So I want to recognize you for your partnership and your commitment, and wish you Godspeed. Thank you. [applause]

Introduction to Congressman Roscoe Bartlett by Charles Manto

CHUCK MANTO: At this time, I welcome to the podium Congressman Roscoe Bartlett. I believe he is here. And I'd like someone to accompany him, if we can. Thank you.

We also expect during the course of the day that Congressman Trent Franks will join us as well. But I just wanted to say a few words about the Congressman.

First of all, I was very pleased to be a member of his district. I had been living in Cumberland, in western Maryland, and working there and also in Annapolis for a number of years. And I hadn't known much about him. I had been the lead of a group of companies that was organizing a strategy to deploy and obtain investment for broadband technology in rural areas. And at the end of that report, we basically said if you're a rural place, you don't have enough customers to attract the leading investment in your area most of the time because you don't have the financial attractiveness for that to happen. And the strategy we came up with basically said we need to use the truth about our remoteness in rural areas to our advantage, which was that we could be a great place for a backup data center or a disaster recovery center. And we basically said we need to be able to convince urban-based governments and private sector to put some of their most special assets in these rural areas, like Cumberland, Maryland, in Allegany County, where the Congressman was represented.

And it was a great strategy. And then we had our final report come out in January 2004. And as I was going around explaining what that strategy was all about, people started asking me, Have you read Congressman Bartlett's report? And I said, What report are you talking about?

So it turns out that during the course from January on, I kept hearing about the Electromagnetic Pulse Commission that Congressman Bartlett help author and lead. And it basically had this very simple message: somebody could take a relatively inexpensive nuclear weapon, a 510 kiloton weapon, stick it on a very inexpensive crude missile, put it on something like a freighter 200 miles offshore, launch it over someplace over Pennsylvania, 80 miles high or something, detonate it, and in nanoseconds impact critical infrastructure from Boston to Atlanta to Chicago, and take out all of those data centers or destruct or damage those data centers that I was advocating for.

And I said, great, I've got all these people ready to come to my rural area, but the rural area's going to be collateral damage. He just undermined everything I was just doing in effect with that single message. And I started looking at this problem and I said this sounds like the one I used to work on in the computer industry during the Cold War.

And this is more of an introduction than you need, but those who are IT people, you'll understand. When you type on your computer, bad guys can sit in the parking lot and intercept those signals; that's what we call hacking today. There's a program for that *(called TEMPEST),* but I never imagined somebody in the van *(with a high-powered microwave)* or somebody with a nuke is going to send EMP signals to us and disrupt our infrastructure. So I had to really become engaged in this. And from that moment I started working on it.

But I discovered Congressman Bartlett, and he was one who was in Congress as a scientist; he's a Doctor Bartlett. He has 20 or so patents. He has been recognized by Congress, by his peers, for leading concern about energy and peak oil issues, and things like that. So it's no surprise that with his great curiosity, intellectual background and experience that he would lead the charge that became the Electromagnetic Pulse Commission, which we will hear from later today, and Dr. Pry and others will talk about that briefly.

But he has a wealth of background. If it weren't for him and his work, we would not be doing what we're doing today. And I just wanted to make certain that first of all we give Congressman Bartlett a round of applause and welcome him to this meeting. [applause]

Congressional Perspective and History of and Report from the New EMP Commission

Congressman Roscoe Bartlett and Dr. Peter Vincent Pry

CONGRESSMAN BARTLETT: What I'd like to do is ask Dr. Peter Pry to join me, and we'll just sit down here and have a conversation because he was involved with this from the very beginning. And we're already a bit behind in our schedule; that's always true here, it's such a tight schedule. Maybe we can tighten the schedule up a little bit and do both of our presentations. Thank you very much for coming. Our microphones are on here? Okay.

I really don't remember how I first learned about EMP, I have no memory of that. But I remember calling Tom Clancy, who was a friend and helped with these type of fundraisers. I thought I was conservative until I met Tom Clancy. [laughter] He was way out there. And I knew he did really good research for his books, and I was curious about EMP, and I knew he had an EMP scenario in one of his books. And so, I called Tom Clancy.

He said, "If you read my book, you know all I know about EMP. Let me refer you to the smartest man hired by the US government." And in his view, that was a Dr. Lowell Wood – he may have been right, by the way, Peter – Dr. Lowell Wood from Lawrence Livermore Laboratory. This was pre-cell phone days; remember the old pagers, the satellites? I paged Lowell Wood, thinking it would go to that satellite in California. He happened to be in Washington, DC, and within an hour he was sitting with me in my office.

I really don't remember which year this was, and he described to me what would happen in the event of an EMP attack on our country. Newt Gingrich was then the almost outgoing Speaker of the House, and we went and briefed Newt on it. You may remember that in his presidential campaign several times he mentioned EMP? I was his campaign advisor on EMP.

I don't remember how long ago. Let's see, Bob Livingston was the incoming Speaker. Bob never quite made it here; he resigned and left. We briefed him, also. He used language I can't repeat here, and he described how much it affected him and how frightened and concerned we ought to be as a nation.

Well, I decided to set up an EMP Commission, and Dr. Peter Pry was a senior staff on the Armed Services Committee. I don't remember, Peter, our first meeting either, but Peter had been involved with EMP in the CIA when he was there. We wanted to set up the EMP Commission, and neither the chairman of the committee, nor the staff director wanted anything to do with us, as you remember.

And they said, "Gee, that would be nice, but we really can't do that because it would require sequential referral. We never do anything in Armed Services that requires sequential referral." What that means is, it has to go to other committees because they also have jurisdiction here. "And we're always the first bill out, so we don't have time to do sequential referrals." I said, Do you know what other committees would be involved?" And they told me, and so I went to the chairman of those committees and I got a letter waiving jurisdiction here. So then I came back and we took this to Peter and they very reluctantly, because they didn't want to do this, they very reluctantly put this in the bill, and the EMP Commission was set up. Peter was the chief staff member there and it run for, what, six years, right?

PETER PRY: Eight.

CONGRESSMAN BARTLETT: Eight years? Okay. Hope it's going to be reestablished now the money is there. And Peter will spend a few minutes here as we discuss this and describe where we are now and what's happening.

There was something that overshadowed the report of the EMP Commission to the committee. Duncan Hunter was the chair. And it was a very perfunctory presentation. What was the event that overshadowed that?

PETER PRY: I'm not sure the thing you're referring to is, was it the Marsh Commission report on report cybersecurity?

CONGRESSMAN BARTLETT: There was some event that overshadowed that, and nobody paid any attention.

PETER PRY: When you and Curt Weldon led the Congressional delegation in Vienna to meet with [4:27]

CONGRESSMAN BARTLETT: Oh, yeah. Okay, that might be worth it to note. What year was that?

PETER PRY: That was April of 1999.

CONGRESSMAN BARTLETT: April of '99.

PETER PRY: Okay, Balkans, and you met with the members of the Russian Duma–

CONGRESSMAN BARTLETT: We did.

PETER PRY: –looking about how to possibly end–

CONGRESSMAN BARTLETT: I'll tell you how to fix that date. It's the date when Slobodan Milosevic turned over the three prisoners, hostages, whatever we called them, to Jesse Jackson. And there was a picture of Jesse Jackson holding hands and standing in a circle, praying with Slobodan Milosevic. And the Secretary of State had a dim view of that.

But we were in Vienna, Austria, in a hotel over there, negotiating the terms that would end the Kosovo conflict. Just a few days later, the GA adopted– Curt Weldon was the chair of our delegation.

By the way, Bernie Sanders was there. And Bernie Sanders and I sat like this in three days of these meetings. And there wasn't any light between us on this. Bernie's a very bright guy, he's a very decent guy; he's just wrong. [laughter]

The thing that probably helped me most as a popular thing was my first degree was in theology, where I learned to love the sinner and hate the sin. So I can like people whose political philosophies I can't support. Nancy Pelosi was and is a good friend. And I've been on several CODELs with Nancy. I said, "Nancy, I really, really admire your commitment, but to all the wrong things?" [laughter] That made it possible for me to like people in spite of what they stood for as a person. She's a very decent person, but where she wants America to go is exactly the wrong place for America to be.

But what happened in that room in Vienna, Austria, the ambassador to this country, Vladimir Lukin, was there. He was then the chair of their equivalent of our [6:33] international relations committee or something like that. And he was very angry. He sat there with his arms crossed looking at the ceiling for three days. He said, "Why did you spit on us?" Clinton had told MG, "We're the big guys now, you really don't count, we'll take care of this thing, the Kosovo thing." And the reality was that the Russians were the only people that the Serbs had any conflict. In fact, Russians told us, Peter, that whatever we decided with Russia that the Serbs would agree to, that turned out to be true.

But after about three days of this, this is what Vladimir Lukin said, "If we really wanted to hurt you, we'd launch an SLBM, we'd detonate a nuclear weapon high above your country and cut down your power grid and your communications for six months or so." And Curt Weldon said, "Did you hear what he said?" Well, of course I heard; I didn't understand a word of it, but Curt did understand Russian. So then the interpreter interpreted and I knew what he had said.

Then the third-ranking communist there, a tall, blond guy, who– what was his name? Alexander Churbanov. And he smiled and said that one weapon wouldn't do it, we'd have some spares. [laughter] I think at that time, about 10,000 spares, is about what they had at that time.

Well, this was quite a remarkable thing, and it got very little play in this country. Because here was a leading Russian saying publicly what they could do with a single weapon over our country. Peter, the reason I brought it up is I have some questions that I don't have any answers to, and maybe you can help me. Here is something that, if it happened, would end life as we know it.

What Vladimir Lukin said is exactly what would happen. It would shut down all of our communications. If you don't have electricity, you don't have anything. The bank has no idea how much money you've got in the bank unless they look at a computer screen. The water won't come out of your faucet, your sewage won't be treated. The financial institutions fall apart, you can't have any healthcare in the absence of this.

Now, I grew up during the Cold War. I just went through my 90th birthday, by the way. I grew up during the Cold War. You could not go into a public building without finding there stacks of pamphlets telling you what you ought to do and how you ought to do it to prepare for this.

And now we have something that's going to end life as we know it. I'm having trouble understanding why– in how many years has it been now since the EMP Commission was set up?

PETER PRY: Since it was established, 16 years.

CONGRESSMAN BARTLETT: Sixteen years. And still, most of the people I talk to know nothing about EMP. I don't understand how we can be in a leadership position and not tell our people that this thing has potential. The probability is not zero. I travel 20 years back and forth to Washington; now I go back and forth to Washington for 19-and-a-half years before I saw the first house on fire. And every house I passed, and there must have been, if you count each house, hundreds of times that I passed, I passed millions of houses. Every one of them, I'll bet you, have fire insurance. Because what you do when there is a low frequency/very high impact event is you buy insurance. A young husband who has kids, and he buys life insurance because if he died, his family would be devastated. Or most people, if their house burned, it would devastate them, so you buy insurance.

Why in the heck haven't we done a similar thing to protect ourselves? Why doesn't every person or country today know nothing about EMP or what they ought to be doing to prepare for it?

PETER PRY: Good question, sure. In part, the government is partially responsible for that, I think. Because the information when I was working on EMP and nuclear issues, nuclear weapons design issues at CIA, all this stuff was equally classified.

CONGRESSMAN BARTLETT: Just one moment, Peter. The CIA, when you were there, the most recent paper on EMP, it was Peter Pry's paper of a decade earlier.

PETER PRY: Yes, that's right. A paper I'd written in 1987, at a briefing, at a hearing we had that was conducted in, I think, 1997, was the date of that hearing. So until the EMP Commission declassified its reports, a lot of the information in its reports in 2004 and 2008, it just wasn't out there in the public sector. And when the hearings that were held with the EMP Commission Reports weren't really well attended. The press didn't pay a lot of attention to it. So it took some years before it started getting into the public domain.

So that's part of it, the fact that the information and the threat had been deeply classified. But even after all of our educational efforts, and let's face it, there hasn't been that many of us engaged – EMP SIG, my organization, basically it's a handful of people who are knowledgeable about this and are going out trying to educate the public, talking to the press.

So you're starting with widespread ignorance. But also there's a great public skepticism because there used to be– I think science is less respected now because there have been so many hoaxes, or controversial positions taken, for example, over climate change, where people will say, "Oh, well, 99% of all the scientists agree climate change is a real phenomenon." And I'm not saying it isn't, but I'm not saying it is though, right? But it's certainly untrue that there's a consensus in the scientific community about climate change. And the average person, I think, is enormously skeptical. If they've been paying attention in their lives, they know that popular science was predicting a new ice age back in the 1970s, and the population bomb and all of that.

And so, when they hear the latest scare story, they don't know whether to think it's whether science fiction they're being told or not. So there's a level of skepticism that people have now about claims about scientific disasters, including in the cyber arena. Remember in the year 2000 there was supposed to be a cyber catastrophe and it didn't happen? So people are kind of used to scientific and– their government calling wolf, and that tends to make it hard to get through the skepticism.

The last I'd say, the fact of the matter is, and I think it's been evidenced in the recent election, there's a tremendous left-wing liberal bias in the press. The press doesn't like to talk about threats to the United States from our foreign adversaries. Whenever they hear especially– they think that that's warmongering, it's going to be another excuse to get America into another unwanted war – "What are we supposed to do? Invade Iran? Invade North Korea?" Their immediate reaction to this is to deny it.

Even though the explanation for the threat is basically bipartisan. For example, you mentioned that Newt Gingrich had raised the threat during his first presidential election. I remember that very well; I was thrilled to hear it. Here's a person for the first time in history during a presidential debate raising the threat of EMP. And you know what happened? The reward that Newt Gingrich got for trying to educate the American people and being so bold as to make this part of his presidential campaign, the very next day the headline, the front page of the *New York Times* had a strictly caustic, critical article about Newt Gingrich had tinfoil on his head and everybody knows that this is a hyped threat that only the crazy right wing believes in.

And I got together a letter that was signed by the nation's leading intelligence and EMP experts, guys like Dr. Bill Graham, who had been President Reagan's science advisor, who is the free world's foremost expert in EMP; he's the chairman of the EMP Commission. Jim Woolsey, Democrat, CIA director. Chris Earmark[?], who had been the chairman of the National Intelligence Council. Lowell Wood. A dozen of our most prestigious strategic thinkers and scientists told the *New York Times*, no, Newt Gingrich is actually correct. And you know, they didn't even publish the letter; they wouldn't publish the letter. So this is the kind of an attitude you encountered in the press.

Last year, in April of 2015, Ambassador Cooper and I published an article in the *Wall Street Journal* about how the Obama administration was spending nearly a billion dollars to harden Cheyenne Mountain against a nuclear hit from North Korea, and were moving critical assets out of Peterson Air Force Base back into NORAD headquarters to protect them against EMP. And *Wall Street Journal* – now, I consider the readers of the *Wall Street Journal*, they're not– they tend to be of bipartisan nature. I think that they tend to be among the more sophisticated and educated and informed people. And half the responses to that letter, in 2015, half the responses to that letter were, "Well, I didn't know EMP was real. I thought this was the stuff of science fiction." So even now we have a huge educational mountain to climb, just convincing people that this is a real phenomenon, that this is something that they really have to worry about.

So I think those are the factors, the fact that this has been relatively new in terms of being unclassified and being available to the public, that you've got a public that's much more skeptical now when they hear that scientists are making claims, because science itself has become so politicized in our society that the average person doesn't know what's true and what isn't.

And then you've got a real problem with press bias that regards any foreign threat as possible warmongering or setting us up for another unnecessary war. And people hadn't even read the EMP Commission reports, for example, and were claiming that the EMP Commission, what all this was about was an attempt to bring back Ronald Reagan's Star Wars.

Now actually, I think Ronald Reagan's so-called Star Wars, Strategic Defense Initiative, should be brought back as a response. But the EMP Commission reports didn't advocate that. What they advocated was a passive defense in protecting the grid, and it went through each of the critical infrastructures to explain how we do it. And so, the press had obviously not even read the EMP Commission reports. But their theory about what this EMP was about was that it was an attempt to get us into another war by having a preemptive war against North Korea or Iran, and an attempt to bring back the unwanted Star Wars program.

CONGRESSMAN BARTLETT: Peter, today there are other threats to the grid that would also bring it down, and even those aren't really publicized. Ted Koppel in his new book, *Lights Out*– I went to visit him down in Virginia and I asked him what he probability was that this wouldn't happen. If you read his book, he claims that if both Russia China thoroughly hacked into our grid at any moment, they could take over the grid and do anything they wished with it, including shutting it down and destroying the grid. And it was only a matter of time before the bad guys get there. And I asked him what are the chances that this won't happen, and he said zero. So this threat to the grid is a very real threat to the grid.

And then the solar storm. There was a big space weather event in 1859. The last really big one that would have taken down the grid is 1921. This happens about once every 100 years, roughly. I think that Lloyds of London has a statistic that there's a 12% probability that within the next decade there will be a solar storm large enough to take down all or part of the grid. Even this hasn't gotten out there. The American people deserve to know that these threats out there, that they can do something to protect themselves.

Right now, if you've read the book *One Second After,* there's a lot of good technical stuff in there, and it's woven into this novel so people will read it. But 90% of the people in this country are dead at the end of a year. That's probably not an unreasonable expectation if this thing happens. And still, why in the heck aren't the American people told that this is so?

On the Armed Services Committee they would come and, "North Korea's not really a threat yet because they don't have an intercontinental missile which will reach us. And Iran, they may be developing the bomb, but they don't have an intercontinental missile which will reach us." And on the record, you can pull it up, I said: Let me see if I can understand. These people may be evil, they are not idiots. They're not going to launch from their soil. They know we have satellites that will see where a response came from. And the response to that would be, they would need to vaporize. A little country like North Korea would be nothing but– we would vaporize the whole country if they did that. Why would they do that and put themselves at that risk when all they have to do is tramp steamer, a Scud launcher which they can buy for $100,000, and any crude nuclear weapon that has an altitude of 180 miles. That would wipe the whole East Coast from Boston to Norfolk and maybe as far west as Chicago. I don't know if that's a mortal wound for a country, but it's close to a mortal wound.

North Korea is launching satellites. What's wrong with putting a nuclear weapon in that satellite? It's a 300-mile orbit and it goes over the United States, when it's over Iowa or Nebraska, detonate the weapon. Don't tell me that we're not now at risk from Iran and North– if Iran doesn't have a nuclear weapon, North Korea would be happy to sell them one.

Why do they keep saying that we're not at risk of these countries until they develop an intercontinental ballistic missile? Peter, why in the heck would they do that, put themselves at risk? We'll certainly know where it came from if we see it. We have, as far as I know, zero defense. We don't have [21:23] any quality or quantity up and down our coast. We're at zero defense against any weapon launched from the south. Am I correct?

PETER PRY: You are correct, sir. And let me, unsolicited from you because you're too modest, I'd just like to point out that people may not be familiar with you and the role you played in this. Since we don't get to see each other that often, and here we are on the stage together after many years of working together, but I'd like to talk a little bit about Mr. Bartlett and history of how we got started. Because we're all here in this room today and there is a national movement now at the grid against EMP and cyber and all the rest. Even the cyber movement owes its origins to Mr. Bartlett, because the EMP issue is what started off the concern about grid security. It was Congressman Roscoe Bartlett who started off a national movement to protect our nation against the existential threat that is EMP.

And we met, I do have the memory, I remember it, February 1995. I had just come over from the CIA, where I had been their senior expert on EMP, and on many other things, on Russian nuclear weapons and strategy. And I had been assigned to the House Armed Services Committee professional staff. There's a professional staff, a personnel staff. There's a collection of technical and strategic and intelligence experts that serve basically all 60 members on the Committee, and they assigned me to work with Mr. Bartlett and Congressman Curt Weldon at the time.
The last thing I had worked on in the CIA was an event on the 25th of January, 1995, which is to this day not well known, but it's the closest we ever came to a nuclear war. Norway has a space program and had launched a meteorological rocket to investigate the aurora borealis. And due to

a clerical error, the Russian general staff hadn't been notified, and they thought this was an attack, frankly, the thought it was an EMP attack.

It's the first time in history we ever had a ballistic missile in flight, and all three chegets on the Russian side – they have three so-called nuclear footballs they can use to execute their forces – and all three of them were executed, all three of them were activated. And two of the cheget-holders, Kolesnikov, who was chief of general staff, and Grachev, who was the defense minister, were yelling at Boris Yeltsin to push the button. Any one of them could push the button independently, but they wanted the president to do it. And they're supposed to push the button when the chegets are activated. And he wouldn't do it.

The whole thing lasted 20 minutes, but we were literally one-pushed-button away from a thermonuclear holocaust. And people over in Congress wanted to know, well, why is that? Why would one meteorological rocket were they willing to push the button, or they came so close to pushing the button? And the answer was, because with one weapon, you can do an EMP attack that would fry their command-and-control, and then it would be followed by a massive first strike by us to take out their bombers and air equipment and the like.

And Mr. Bartlett was the only member of Congress who knew about EMP. And the EMP Commission that we all give so much credit to, before there was an EMP Commission, which wasn't established until 2001, there were five years of hearings. This man wandered in the wilderness, and he started off almost alone, and put up with a lot of ridicule and criticism from people for being concerned about EMP and the idea that, hey, the electric grid is critical to our national security. But slowly, by bringing in the facts and having the first open hearings in history on the EMP threat, we were able to educate members of Congress and bring them along.

And I think the decisive moment came in the incident that you described in Vienna, when you had a bipartisan Congressional delegation meeting with the Russian Duma about how do we end the Balkans war. Don't forget, the Balkans was the first time NATO had actually gone to war, and we were bombing Serbia, which was a longstanding historical Russian ally. World War I got started because of Russia's concerns about defending Serbia from aggression.

A few months before that meeting, we had been told by the Clinton administration, by the Marsh Commission, which was a commission on cybersecurity, not to worry about an EMP attack. Mr. Bartlett had asked General Marsh, "Well, what about a nuclear EMP attack in addition to the cyberattack?" And Marsh judged that, "Look, the Cold War is over. We don't have to worry about Russia anymore. And only the major actors, like Russia or China, can even think about an EMP attack."

And shortly after that, this bipartisan Congressional delegation, meeting with the Russians, was threatened to its face with a nuclear EMP attack on the United States by Russia. And when they got back, the EMP Commission got established, and we started having our first meetings. Because it was clear that the defense and intelligence agencies, departments and agencies of the Clinton administration had dropped the ball, that they weren't paying attention to this threat, that this was something real.

And indeed, when the Commission met later, we found that just as we had suspected, in the hearings that we had for those five years, that it wasn't just the Russians and the Chinese. The North Koreans, the Iranians, the Iraqis – this was before Iraq had been conquered – that they knew about EMP and incorporated it into their plans. And indeed, as we'll talk later, EMP and cyber and physical sabotage are actually all part of their plan, that they basically have developed this new way of warfare, in which EMP attack is the ace. I mean, it's the most effective, most devastating thing if they apply to use it in coordination with cyber and physical sabotage attacks. And not only do not people know about EMP, but even more importantly they don't know about this new way of warfare, sort of a blitzkrieg in cyberspace that is the basis for military thinking in Russia, China, North Korea and Iran, and even amongst some terrorist groups, if they can get their hands on the bomb.

We are spending our time here in this country arguing about, well, what deserves higher priority? Cyber is more likely than EMP, isn't it? And then the Green Beret types, the SEALs will tell you, Well, a sabotage attack on the grid is even easier to do, is more likely than cyber, isn't it? So we need to put all our resources into physical protection. No, it has to be cyber. No, listen, EMP, what about that solar storm, isn't that what we need to be preparing for? All [28:20] It's all of the above.

Anyway, I go back to you, sir.

CONGRESSMAN BARTLETT: Peter, why don't you tell us the additional things you were going to talk about and maybe get our meeting back on schedule. Go ahead with your 11:00 that you were going to say there.

PETER PRY: Well, you're the commander-in-chief so I'll– [laughter] I'll try to [28:49] so George can take over that handle and have the time. Chuck wanted me to talk about where is the EMP Commission today and what are developments that are going on today. Well, the EMP Commission, I remember when we delivered our final report. And by the way, the House Armed Services Committee, the House, didn't want the EMP Commission to end. In fact, nobody wanted the EMP Commission to end. The Commission was supposed to continue past 2008. The idea was that it was to continue until the grid and the other critical infrastructures were protected.

The problem is, the Defense Department didn't want to pay for it. The Defense Department said, "Well, the EMP Commission has done good service. It's enlightened us, and we agree with them about this EMP threat for military forces. But now the Commission is spending all its time looking at protecting the electric grid, and the telecommunication system, and the water and food supply systems and the like, these civilian critical infrastructures. And this isn't our department. This falls to the Department of Homeland Security. So we recommend that the EMP Commission continue, but be reauthorized under Department of Homeland Security."

And we tried to do that, but unfortunately – this is kind of inside baseball stuff – because of sequential referral that Dr. Bartlett referred to, there has never been an authorization bill for the Department of Homeland Security that has ever passed Congress. Because the various committees, we never dissolved the committee chairmanships. There's still a committee on transportation, there's committees on electric power, banking and finance, and on and on. These chairmen, they always stop the authorization bill because they don't want to see jurisdiction for their committee. So the authorization bill for the Department of Homeland Security never passes, and we've had to reauthorize the Commission through the authorization bill.

So we tried and tried and weren't able to do it. As a stopgap measure I formed my EMP Task Force which is like one of the lowest life forms of [simultaneous conversation] Congressional advisory board. It's nothing like a commission. It doesn't have power, it doesn't have resources.

But finally in frustration, and because the Commission recommendations were being ignored for almost a decade, the Congress of 2015 brought the EMP Commission back, the original commission, and we were reestablished in November 2015, around Thanksgiving.

And we've done what we can, but to give you a sense of how bad, how perilous our situation is as a society, for the past year, ever since we were reestablished, Obama's Department of Defense has slowrolled the Commission. They've been sitting on our money. We haven't gotten a penny of the two million dollars that was appropriated to us. They've finally gotten around, not all of our staff have security clearances yet, but they've been dragging their tails on getting security clearances.

So basically, we've been established legally as a commission. We exist legally as a commission. But we have no resources to do any work. We don't have security clearances so that we can go and look at classified information.

The President-elect of the United States, a couple weeks ago, maybe it was last week, directed the Joint Chiefs of Staff. He said, "When I become President, I'm going to direct the Direct Joint Chiefs of Staff to come up with a comprehensive plan to protect the electric gird and other critical infrastructures from cyberattack and all other threats."

You'd think with such direction that that made a difference, right? And that overnight we would see the Department of Defense be forthcoming since the next administration takes this stuff so seriously. But that hasn't happened yet.

Nonetheless, we are continuing to do what we can on our own, on a pro bono basis. We will not be thwarted from delivering a report. The Commission's going to deliver a report, and it's going to be cutting-edge analysis. I did a sample, a book that was published. It's not a Commission product, but it's the kind of analysis that I think Dr. [33:01] would agree we should go forward with. I just published a book for the Center for Security Policy called, *The Long Sunday: Nuclear EMP Attack Scenarios*. And for the first time, it actually shows, we've known EMP can be used not just against the United States, but against our foreign allies. And it goes through theater, Asia, Europe, North America, the Middle East, and talks about the various scenarios, how EMP attacks could be used to advanced geostrategic interests of our adversaries. It talks about using that in combination with cyberattacks and sabotage within those theaters.

And this is cost me thousands of dollars of my own money, okay? There's two million dollars sitting for the EMP Commission, all right? But out of my own pocket I paid to do this cutting-edge analysis. What we've done is shown what the consequences of an EMP attack would be in terms of what I call blackout zones. Most of the focus, when people think of a nuclear EMP attack, they think of the EMP fields. But what isn't understood is that damage occurs way beyond the field. The electric grid that's exposed to the field oftentimes will go outside the field. And so, the total blackout zone actually propagates far beyond where the EMP field is.

That's why you could take a balloon and put a one-kiloton nuclear artillery shell in it, put it over the East Coast of the United States. It will only have an effective radius of 600 kilometers; that's not covering any– it's equivalent to maybe five states or something like that. But the effect will be to black out the whole Eastern grid, which would basically end us as a civilization, because we can't survive without the Eastern grid; 75% of our electricity is generated in the Eastern grid, and most of our population lives there.

So understanding this concept of how does the damage from EMP propagate beyond an EMP field is kind of a new way of thinking about war. And this books shows, in theater after theater, including in the United States, how that would work and how it would fold into the geopolitical and geostrategic interests of our various adversaries and various theaters of operation.

Now, the other thing that I would say, to give you a sense of how bad we are, despite all these years of the EMP Commission, and all the efforts of groups like this to educate the public and try to move the problem forward, is that while the EMP Commission is being stonewalled, the Obama administration's Department of Energy is going forward, gung-ho, with a project in cooperation with EPRI. That's the Electric– that's one of the organizations, the think tank for NERC, the North American Electric Reliability Corporation, a private sector corporation, that has been denying EMP threats, cyberthreats, physical sabotage threats, and trying to reassure the public and the Congress that everything is all right, they're on top of the problem, these threats aren't really all that serious, the solar storms are greatly exaggerated. That has been EPRI's position. They've produced junk science that we have over and over again torn apart at technical conferences and demonstrated it's false analysis.

We've been given two million dollars that is sitting in the Department of Defense that hasn't been available to the EMP Commission. While we've been being slowrolled with one hand, on the other hand the Department of Energy and the electric power industry, basically, their lobbyists in EPRI, are spending $26 million to try to produce a report that's going to argue, well, the nuclear EMP threat isn't all that bad either. Just like the solar storm threat isn't that bad, the nuclear EMP threat isn't all that severe either. Because the first line of defense for these guys is always going to try to deny the threat. They're trying to deny that cyberthreats are all that bad or that they're on top of the problem.

And the usual suspects are in charge of writing this. It's the same team that back in 2012 signed on to the NERC report that said, Oh, the possibility of a solar super storm blacking out the grid is greatly exaggerated. And that we would be able to, they actually claimed that they would be able to bring the grid back in 24 hours after the event, [37:21] event without doing anything. And fortunately we tore them apart in a technical conference to prove that wasn't true.

NERC ordered them to come up with a standard, okay, called the EMP standard, protecting against solar storms. And since NERC can't define what the standard is, they've come up with this standard that's in effect a hollow standard that will not protect the grid.

So we're really no better off right now. I'm seeing in some respects we're worse off than we were when the EMP Commission started. Because not only have we gotten [37:59] infrastructures against the spectrum of threats, but we're now in a position where we have NERC, EPRI, other parts of industry propagandizing the American people who don't know who to believe. They're not scientific experts and they've heard so much. But propagandizing American people and policymakers not to worry, there's nothing to worry about, Ted Koppel doesn't know what he's talking about, the EMP Commission doesn't know what they're talking about. And really, the utilities are the real experts on all of these matters, when in fact they have no expertise on these matters. They have no expertise on EMP or cyber or physical sabotage.

So in respects, as to where we are today, we are worse off because we've actually– the Commission that Congressman Bartlett established, it was the tip of the spear for protecting our nation and Western civilization against EMP threat. It's being stonewalled with one hand, and gross propagandists and people who don't know what they're talking about are being given $26 million to produce a report I think you'll probably be reading about in the newspapers pretty soon. It's going to say, Oh, new study that comes out from DoE and EPRI that says, We're on top of this nuclear threat, EMP threat and maybe it isn't such a serious threat. Or more likely what they'll argue, they'll say, they like to use the phrase risk analysis. Which is trying to make a virtue of a vice. The essence of risk analysis is, well, what is the thing that's mostly likely to happen? Well, squirrels are much more likely to happen as a threat to the grid than EMP or cyberattacks. It gives little attention to the consequences and ignores the fact that we're very bad at analyzing likelihoods and probabilities.

One thing I know, my original training was as a historian, and history tends to make fools of us all. We're really bad at trying to anticipate what is the likelihood of threat. When we've got two North Korean satellites orbiting over us on the optimum trajectory to evade our national missile defenses and doing an EMP attack on this country, I think the idea that that is a low likelihood threat is wrong. We know for sure the solar storm threat is inevitable.

Anyway, I'm being told by Chuck that I'm over my time already. So thank you so much for hearing me out. [applause]

Congressman Bartlett had a closing remark to make, and plus we have a presentation to him as well. So Congressman Bartlett.

CONGRESSMAN BARTLETT: How many of you have heard of Yamantau Mountain? Wow. This is very disquieting. Let me take a very brief time to tell you what it is. It is the largest nuclear-secure underground facility in the world. It started in the days of Brezhnev, and it's in Yamantau Mountain in the Urals, about 600 miles almost due east of Moscow. And it is huge, it is as big as inside the Washington Beltway. There are train tracks for trains running in two different directions. And we have no idea why they're building it. It's still there. The last brief I had, Peter, said that what the Russians are going to do is take the cream of their civilization there so that they re-populate after a catastrophic event had wiped out essentially everybody else.

Joel Skousen, who believes that– I don't think this is true. I think there are plenty of other places that [41:36] biological and chemical weapons which they say they're not doing, which I think they are doing. But [41:42], whatever they are doing in Yamantau Mountain is enormously disquieting. There is no good explanation of what they're doing in Yamantau.

If you don't know something about it, read up on it. What it means is that Russia expects that there will be a cataclysmic event which will destroy most of their people and they'll put the cream of their civilization in Yamantau Mountain so that they can re-populate after this cataclysmic event.

Couldn't that be disquieting? I think so.

Thank you very much. [applause]

PRESENTATION OF *POWERING THROUGH* TO CONGRESSMAN BARTLETT BY EDITORS AND AUTHORS

CHUCK MANTO: Last year, many of these people have been working every week to tackle the problem of high-impact threats, now that we realize we're not ready for these high impact events on our society. We've started the work of bringing top experts together to just scratch the surface for how to do comprehensive planning, whether you're a company or a jurisdiction or a community, to be able to work through this. And they did a great job. Let me just hold this up for the audience. [*As a result,*] We have a new book called *Powering Through*. It's the first comprehensive planning guide for communities and government agencies to be able to look at this, drawing from all the work, such as the EMP Commission and others.

And we've had endorsements from folks ranging from Ted Koppel to the guy who wrote *Chicken Soup for the Soul*, and people in politics as well as people in technology. We have wonderful statements and involvement from government entities, including DHS who can't formally endorse a book, but they can talk about the significance of the issue and why we should really begin to work together to plan for things that we normally hope we don't have to bother with; we'll let somebody else do it.

Well, we've come to the time when in fact Congress and the White House, with their new space weather strategy, actually calls on, not only the whole of government, but the whole of community to begin to plan for things like this under the rubric of extreme space funding. And that's one stalwart position that has come out in the last year which has made a substantial change in what we think is the right direction.

But the group here has already pulled together this first effort at such a planning guide, and I would like to have all of you who are standing here – Bill, Steve and Mary – present you with a copy of this in thankfulness to you and your work, Congressman Bartlett.

CONGRESSMAN BARTLETT: Thank you very much.

[applause]

PSA Program, DHS and ties to EMP SIG

PSA Director Scott Breor Introducing Assistant Secretary Durkovich

CHUCK MANTO: So now what I'd like to do is introduce Protective Security Advisors programs director. He can tell us more about his official role, which is beyond just that. But I wanted to highlight that. Many of you know Scott Breor, who is going to come up for a moment. And you learned last year when we were introducing and he spoke to us that we were so lucky to have him for a couple of reasons.

Number one, as you know, InfraGard works with the private sector to try to merge private sector and public sector together. But there's this great activity in the Department of Homeland Security that Scott leads, which takes seasoned professionals and brings them at senior positions with the senior level positions at DHS. There's now over 100 of them standard throughout the country who integrate very closely with the private sector and any of the issues that really concern them, and are beyond the ability of any one industry to work on their own.

And so, we thought this would be a great opportunity to get their program and their top leaders to integrate with what we're doing here through the InfraGard EMP SIG. And he's going to mention that briefly as he introduces our next speaker. And he will speak more about it later in the morning because we're shuffling our schedule around a little bit.

But I just wanted to mention one thing about him. In addition to being a fighter pilot– and he's going to tell you his handle, if he wants to. They all have handles. General Newman's here; he was a pilot, he's got a handle. But anyway, he did a master's degree in physics in the '80s, and of all the things he could have studied it was physics. And guess what his thesis was on? Do any of you out there remember what it is from last year? Physics thesis, do I see any hands, remember what his thesis was? It was the impact of extreme space weather on electric power grids! And here he's now working at DHS, integrating their work, protecting infrastructure, in the very field in which we're talking.

So I think it's very appropriate that he come up now and introduce our next speaker. Thank you very much. Scott?

SCOTT BREOR: Thanks, Chuck. And coming up here right now on this stage really reinforces how much we support, not only InfraGard as a whole, but the EMP SIG. Because when I showed up this morning, Chuck didn't come and ask me to introduce my boss; Chuck directed me to. [laughter] When I met Chuck over a year ago, it's been sort of our relationship, plus being fortunate to work with Mary, Kelly and many others. It's really been truly awesome.

I'm fortunate with the job that I have because I work for an individual. She may not remember me, but I first met her back around '04 when I retired from the Navy and came over to DHS. And one of the efforts I was trying to support the Department on was sort of a doctrine for DHS. And right there, that's one of my sweet spots because my last tour was in the Pentagon and I was the lead for a Navy joint doctrine at that time. So I was very passionate about that, and that's where I first met Caitlin.

And then, in '07, that interaction became professional because she joined the Department as chief of staff for the National Protection and Programs Directorate. And if you think about where that directorate was at that time – which has cybersecurity, communication, develops infrastructure protection, Federal Protective Service; it also had an entity, US visits, so it had biometrics at the Department – we didn't have the National Cyber and Communications Integration Center at that time.

So these are all key issues for DHS that Caitlin worked through and led many of as the chief of staff. Fortunately for me, she came over as the assistant secretary for the Office of Infrastructure Protection a little over four years ago. And I say fortunate for me because there's nothing more enjoyable than coming in to work and knowing that the passion that drives your boss is just to ensure that communities are safe and secure. And that's what we're all about within our office, and that's what all of you are about. And that's why you're sitting here.

I can't say [2:39] but she is a Blue Devil. And from that, I'd like to introduce the Assistant Secretary for the Office of Infrastructure Protection, Caitlin Durkovich. [applause]

Recent Developments in DHS and Space Weather Executive Order

Hon. Caitlin A. Durkovich, DHS Assistant Secretary

Introduction by Director Scott Breor

SCOTT BREOR: Thanks, Chuck. And coming up here right now on this stage really reinforces how much we support, not only InfraGard as a whole, but the EMP SIG. Because when I showed up this morning, Chuck didn't come and ask me to introduce my boss; Chuck directed me to. [laughter] When I met Chuck over a year ago, it's been sort of our relationship, plus being fortunate to work with Mary, Kelly and many others. It's really been truly awesome.

I'm fortunate with the job that I have because I work for an individual. She may not remember me, but I first met her back around '04 when I retired from the Navy and came over to DHS. And one of the efforts I was trying to support the Department on was sort of a doctrine for DHS. And right there, that's one of my sweet spots because my last tour was in the Pentagon and I was the lead for a Navy joint doctrine at that time. So I was very passionate about that, and that's where I first met Caitlin.

And then, in '07, that interaction became professional because she joined the Department as chief of staff for the National Protection and Programs Directorate. And if you think about where that directorate was at that time – which has cybersecurity, communication, develops infrastructure protection, Federal Protective Service; it also had an entity, US visits, so it had biometrics at the Department – we didn't have the National Cyber and Communications Integration Center at that time.

So these are all key issues for DHS that Caitlin worked through and led many of as the chief of staff. Fortunately for me, she came over as the assistant secretary for the Office of Infrastructure Protection a little over four years ago. And I say fortunate for me because there's nothing more enjoyable than coming in to work and knowing that the passion that drives your boss is just to ensure that communities are safe and secure. And that's what we're all about within our office, and that's what all of you are about. And that's why you're sitting here.

I can't say [2:39] but she is a Blue Devil. And from that, I'd like to introduce the Assistant Secretary for the Office of Infrastructure Protection, Caitlin Durkovich. [applause]

CAITLIN DURKOVICH: Thank you very much, Scott, for that kind introduction. Yes, I am a Blue Devil; loved or hated, but never ignored.

It is a privilege to be here this morning. I have proudly served– and Scott got the dates slightly wrong. I came to the Department in 2009 as part of the Obama administration. And I will tell you, I have proudly served our President for the last seven-and-a-half years and am equally as proud of the accomplishments that we have made over that course of time.

And I want to spend some time later in my remarks speaking about it, but I will, just to the passion that Scott mentioned, tell you – and I have said this publicly and it's ended up in articles even – I really do think I have the greatest job in the federal government, and potentially the greatest job ever, because every day I get to come to work and work with owners and operators of our critical infrastructure in managing what has really become a very complex and dynamic risk environment. No one day is the same when you look at the range of issues that we deal with. But, equally important, the people, my colleagues that I go to work with every day, are so passionate and committed and dedicated and selfless to executing and implementing our mission.

And I've known that January 20th is coming for about seven-and-a-half years, and I have to tell you it's bittersweet. I'm not really ready to walk out the door. But I do take a lot of solace in the fact that I have a phenomenal leadership team in place in Infrastructure Protection, and I know that I'm turning my baby over to folks like Scott and it will be in good hands. So thank you very much, Scott.

I also want to thank Chuck and Mary and Kelly really for the fantastic partnership that we have. We've come so far over the last several years. And it is really a joy to be able to participate in these types of events.

I also want to thank the members of the InfraGard NCR chapter, who have brought to life the intents and themes at this year's national EMP SIG sessions at the Dupont Summit. The work that InfraGard does, not only across the country, but the NCR chapter, the EMP SIG, and also the International Council on Systems Engineering, all the work that you've done leading up to today's event, including hosting what transpired yesterday, the review of the *Powering Through* planning guide, which is intended to help us think about how we are prepared and plan for these types of events, is so critical. So thank you.

I know that you all have day jobs, and am grateful that you come together to focus on what I call these types of kind of unbounded events. And I'll explain a little bit more about that in a bit.

But I can't tell you how important sessions like today are because at the end of the day we are living, as I mentioned, in what is a rapidly evolving and equally dynamic risk environment. And when you look back 12 years ago, as we started to stand up the department and what we were really focused on at the end of the day, which was largely counterterrorism and mitigating the potential threat of a terrorist-directed group, al Qaeda, and how we evolved, I think that really speaks to the dynamic risk environment that we live in.

Yes, we are still concerned about al Qaeda, but we are also equally concerned about this terrorist-inspired environment that we live in – the rise of ISIL, their use of social media and the Internet to recruit, to train, to arm, and to direct our adversaries, those who have grievances against the West. And equally important to say you don't have to travel overseas to fight with us anymore; you can stay here at home. The battlefield can be at home and you can pick, most recently, a car and a knife at a location of your liking and perpetrate your attack. It's a very real world that we are living in right now. We've seen it in Orlando, we saw it at Ohio State, at San Bernardino, and all around the country.

As Scott alluded to, we have seen the rise of cyberthreats, vandals, thugs, thieves, saboteurs, all who understand the importance of our critical infrastructure and who are looking to gain footholds to steal our intellectual property, and who are using cyberspace to perpetrate their attacks.

We have an infrastructure that is aging and failing. Much of our nation's critical infrastructure was built nearly 100 years ago, some of it more than 150 years ago. I guarantee you that the people who built that infrastructure never could have imagined the world that we live in today or account for the population shifts and the urbanization and the incredible ecosystem that we have built that powers what we do every day.

We are dealing with the emergence of technology in things like position, navigation and timing that have become ubiquitous across our critical infrastructures. And yet, with PNT we rely on a single source, the global positioning system, which every day is increasingly at risk. And we have to think about how we create resiliency and redundancy in GPS.

And then equally important, these bigger, unbounded events, things that frankly I never imagined coming into the Department – the risk of solar storms, GMD, and the reason that we're here today, electromagnetic pulses.

I think all of this is made more complex by the fact that we are increasingly interconnected and interdependent when you think about this ecosystem that is critical infrastructure and how much our water systems rely on the Internet, rely on electricity, rely on all of these important lifeline sectors. And you can do through each sector and point to the critical dependencies and other lifelines.

Again, we've built a very sophisticated and complex environment where a disruption, not only in one sector but at one particular asset, can have cascading impacts across multiple sectors. And so, when you take an event – whether it is a Cascadia subduction zone earthquake, whether it is a space weather event, or whether it is an EMP disruption, these kind of massive regional or national type of events – you begin to understand and realize the types of impacts and consequences that we have and why we have to come together as a community – private sector, the federal government, state and local governments and other entities – not only to understand the risks that we're dealing with, but to think through how are we best prepared.

And that's really, at the end of the day, the mission of the Office of Infrastructure Protection, is to help owners and operators of critical infrastructure working with our state and local governments to manage this very complex risk environment. We're not in the business of telling people how to do things; we're in the business of supporting them. We're a convener and a coordinator, where we bring people together, ask difficult questions, share information and, again, work towards ensuring that we are better postured to handle both the routine, but the unbounded.

I want to talk a little bit about what the federal government is doing to address EMP threats and hazards. And I do think that this is really illustrative of the role that the federal government plays in working with owners and operators of critical infrastructure to ensure the security and resilience of this ecosystem.

Chuck mentioned some of the work that we have done on space weather. Again, this has been one of the most surprising, but most gratifying efforts that I have participated in and serving as the co-chair of the Space Weather Operations, Research and Mitigation, or SWORM, Task Force. And I continue, at least I will for the next 50-some days, in overseeing the implementation of the Space Weather Action Plan.

But really, at the end of the day, we undertook this effort to highlight and to continue to drive forward on the important impacts that space weather research and mitigation has on the benefits associated with EMP, including improved electric grid modeling, contingency planning, and the hardening of components.

And I want to tell you, one of the most remarkable things that we did is this task force. And I think that this is different than other task forces that have been convened. We made sure owners and operators of critical infrastructure were part of this effort. We didn't gather as a federal community and come up with a strategy. Through the FRNs, through a number of meetings that we did, we brought the private sector in, both to talk about this particular hazard, but to better understand how it would impact their infrastructures, their systems and their networks, and what we needed to be doing as a federal government to ensure the resiliency in the event of this type of event.

Equally important, our Office of Cyber and Infrastructure Analysis is partnering with the Department of Energy, and has funded a project at the Los Alamos National Lab and the Sandia National Lab to analyze the hazard environments, the impacts and consequences of different sources of EMP and GMD on the electric power infrastructure, and to use those methods to determine EMP and GMD events of concern.

The study's going to be carried out in four phases, each of which will provide higher levels of analytic fidelity. It focuses on those EMP and GMD sources, and the events that create significant consequences, or whose consequences are sufficiently uncertain and help guide us in doing additional study.

We are also working very closely with DoE and its work to support the development of an EMP strategy in concert with the Electric Power Research Institute. That's EPRI that was alluded to earlier.

We also partnering with other interagency partners to complete work on the Strategic National Risk Assessment as part of the National Preparedness System Refresh, and unclassified summaries of the national risk assessment will be released, along with other National Preparedness System documents upon review in federal concurrence.

In tandem with each effort, EPRI, which was mentioned earlier, is initiating a study of EMP vulnerabilities that includes the participation of over 35 utilities, with $5.4 million in federal

R&D funding. This effort and associated funding is important in helping us to better understand the consequences of an EMP attack.

And this is where I do want to pause for a minute because I think the focus that our communities have given on these lower probability, but high impact, unbounded events has really changed how we are engaging and participating in these efforts. And I think it is worth pointing out that both the federal government, but some of our lifeline sectors do take these threats very seriously.

Three days ago, we convened what we call the Electric Sector Coordinating Council. This brings together very senior level policymakers in government with 40 CEOs of the largest electric utilities, public power associations and world cooperatives, to take on the hard issues. We come together three times a year. We've got a work plan that addresses cybersecurity, that addresses information sharing, and that addresses these unbounded threats.

So I don't want you to leave here today thinking that we are trying to sweep this under the rug and that we are not taking this seriously. We are very engaged in these efforts. We are working very closely with our industry partners, both to ensure that these threats are well understood, that the consequences are well understood, and that we have a plan that begins to address how we will respond in the event that one of these things happens.

Scott is going to talk a little bit more about our PSA program and the support that we provide to the owner-and-operator community, both in terms of helping understand this complex risk environment, but the tools that we bring. One of the cornerstones of what we do at the Office of Infrastructure Protection is something called a Regional Resilience Assessment Program, where we look at a particular industry and its dependency on lifeline sectors and the threats and hazards that can impact it and how we come together as a community to ensure the resilience of that industry or that particular sector.

We are doing a lot in the electric industry. This is both because of our partnership with industry, but because of the number of both events that we are seeing in cyberspace, because of what we saw out in California and the attack on the Metcalf substation, but also because of these types of unbounded events.

And then, I think finally it is important to note that we also do take very seriously the work that Congress and GAO has done on this. And in fact, in April of 2016, GAO completed an audit of federal actions around electromagnetic risk. There were six recommendations, four of which are DHS's, and I will tell you again that we take very seriously our commitment to implement those recommendations.

So if there is anything I leave you with today, I hope it is a sense of the commitment of the federal government, of the Department of Homeland Security, and of the Office of Infrastructure Protection to continue to work with our partners, both to manage many of the day-to-day risks that I talk about, but again these lower frequency/higher impact, unbounded types of events. And I think it is important, again, the partnership that we bring together from all parts of government and from all parts of the private sector.

I had an opportunity yesterday to talk to some incoming elected officials and spent much of the time talking about this dynamic threat environment, but also talk about in an election cycle where we have had a lot of conversation about the importance of modernizing our infrastructure or reinvesting in our infrastructure, recognizing that we've come to the life span. The other thing that I think that we need to continue to push is that as we reinvest, as we modernize, as we retrofit, we need to make sure that we are accounting for the world that we know today, but the world that may exist not only in three years, but ten years, in 60 years.

And so, we have to have the security people sitting next to the architects, to the builders and … And so, as we engage in this conversation about the need to modernize our infrastructure, this has to be a constant drum beat. And that is my ask of you. As you interact with members of Congress and others who are thinking about this, make sure that they understand the importance of putting the security folks next to the builders and the architects.

So again, thank you very much for your work on this effort. I'm delighted that Scott gets to follow me and tell you a little bit more about how we're actually out there in the states working with owners and operators and state and locals to help them understand this complex environment, the value that PSA brings to this fight. But I am, again, grateful for the opportunity to address you, to have worked with you for the last seven years. And as I've said many times over the last couple of weeks, it's not really goodbye because, as I've learned in this world, it is a small world, and I know that our paths will continue to cross.

So with that, do you want me to take questions, or are we trying to get back on time? Okay, we're trying to get back on time. Thank you very much. [applause]

PSA Program and Ties to EMP SIG

PSA Director Scott Breor

SCOTT BREOR: And earlier, Chuck mentioned the thesis that I had to work on. What was most impressive– this was back in the early 1980s, so I had to do that on a Brother typewriter that had the single-line memory. And I was using data from the Pioneer satellites, and I actually had to come up with the units and map that myself because if you wanted to– I have a picture of, say, 1000 words, so I needed to come up with those diagrams that truly would catch the reader, and hopefully somebody read my thesis at one point.

So it was to draw those myself, and then run into a library, remembering that I've got the nickels so I could use the copy machine, because I wanted to make copies of my sketches, so then that way it would look professionally done. And then take that copy, put it into the Brother typewriter and then type around it so it actually looked like an important document. So in today's day and age, with computers and everything that kids can do, it truly is amazing.

I'm fortunate; Chuck always gives me a bit of time to speak about the program that I get to work for. I don't consider myself to be the leader of it; I'm just there to support and ensure that our protective security advisors have the resources and tools that they need to best do their job.

We have 102 protective security advisors across the US, and one in Puerto Rico. So if you do the math, you can see that we have a couple in the larger metropolitan areas. And I like to tell people that they are just there to advise, train and assist the local community. And that pertains to whatever that local community believes is a threat or hazard that they might face.

So I will speak to a little bit about why I really enjoy this engagement, because I do get these books from Chuck, and I ensure that they get out to the regions because I know if this hazard or threat comes up, that I want my protective security advisors there to be able to speak to that threat, and then be able to best support, whether it's an owner/operator or whether it's the community, come up with adoptions that might reduce vulnerabilities within that area.

Specifically to the missions that protective security advisors support: One, incident response. It's critical for them to respond to incidents. They typically go to the state emergency operations center or to, if it's a staff-backed[?] at that level, then the [3:14] field office. And they are there to help with the prioritization and restoration of critical infrastructure.

Another thing that they do is vulnerability assessments. I know that a lot of people don't necessarily want to have assessments done to their facilities. But I still advocate that they reach out to their protective security advisor, because then there's information that we would like to have. Between the department and that owner/operator, that information is protected, that information pertaining to dependencies and interdependencies that that facility might have. So when they do have to respond to an incident, that they are better informed of those structures and what those structures require to get back up and running sooner rather than later.

So if you haven't reached out to your protective security advisor, I strongly recommend that you do.

The other mission that they support is special events. Think about the Super Bowl, Daytona 500, a lot of these large public gathering events, part of the city, especially now with the Super Bowl where they have the [4:29] experience. It's typically not just in the city, but it can include a region, depending on where they have that [4:37] experience. So needless to say, with all that security, with all those people, there's a lot of concerns around the critical infrastructure lifeline sectors within those areas. So our protective security advisors work with the planning committees, both federal and state, to ensure that protective measures are addressed as we work up special events.

The last two areas that the PSAs work, and this is where I really see strong leverage between our program, InfraGard and specifically EMP SIG, one is the assessments. They do do assessments at a facility. So they can go around the facility; we have the tools to help them do that. Caitlin mentioned the Regional Resiliency Assessment Program. That is a program that our PSAs are involved with, that our office leads. And they look at what I refer to as a larger geographical area; geographical not being necessarily confined to just this land mass. So it could be a system. We've done system analysis of public works, wastewater for a state, for example.

And why it's important for us to work together on this is because we all have limited resources, so as we initiate these regional resiliency assessment programs, there's opportunities there to scope those projects to include analysis or work that you might need with whatever project you are working on. A good example of that is recently we did Ashmore[?] data center. When we originally scoped that, it was just based on power outage. But working with Chuck and his folks, we brought into that the idea of an EMP. So I think that's a good example, and I think that's the example of how we need to work going forward.

In '17, what we're looking to do for the PSA program, right now we're executing what I consider Caitlin's vision. Right now our program, like I say, we have 102 protective security advisors, and they are primarily supported from folks here in the national Capitol region. We are shifting to a regional service delivery model. So we are standing up offices within the ten federal regions. Within those offices, we will have critical infrastructure specialists, ops analysts, and some other folks. And to me, that's going to increase the bandwidth of the protective security advisor and give them better reachback, because that reachback now is within their region, and those CI specialists and ops analysts can focus on the concerns and issues of that region and be better prepared to support the PSAs. To me, that's a far better way of executing our program than trying to service all the PSAs from inside the national Capitol region.

So I'm embracing Caitlin's vision for that. We're in the process of executing that right now. We have the new regional directors in place. So I imagine in about a little over a year they'll have more capability to better support the InfraGard chapters that are within their region, and also the crosscutting InfraGard areas.

And last, we recognize, we have all our PSA protective security advisors trained in cybersecurity, but we know it's a major concern of the states. We do follow the state preparedness reports. When they say that they're worried about an active shooter, [8:41] cybersecurity, then we ensure that those are three areas that we're going to focus on.

So we got together with cybersecurity communications, and using the National Initiative for Cybersecurity Education, the NICE framework, we have created new baseline training for our PSAs. And then with that, various levels that they can work through so that they can augment the cybersecurity advisors that the Department has in the regions to do some of their cyber vulnerability assessments.

We're very excited about this. We don't see the PSAs becoming cybersecurity advisors, but we see some of them who have been trained to a level that they will be a [9:23] for the cybersecurity advisors.

Specifically, with Chuck's ask, PSA engagement with EMP SIG, what we are doing. First of all, when you read through *Powering Through*, and when you open it up and you see a letter from my boss, and that letter recognizing the issue, and then when I page further along and I see language around National Preparedness Framework, PPD-21, National Infrastructure Protection Plan, that doesn't tell me that it's a matter of should the PSAs support EMP SIG and other InfraGard scenarios of concern. It's how better can we do that. And that's our approach each and every year.

This past year, we established a memorandum of understanding with the National Members Alliance that really embraces and codifies the collaboration that we're going to have going forward. And right now we're working with the Office of the Private Sector that Director Brekke represents to link up the 56 field agents that are out there to support InfraGard, from the FBI, with our protective security advisor cadre. And I've also, with this regionalization effort where we're moving to a regional service delivery model, now all those new regional directors report directly to me.

So in their '17 performance plan, I put in there for them to strengthen and continue to support their work for the InfraGard chapters. I feel that's important. I feel the chapters and us do great work, but I think we can do better work as we work together.

So that is really the update for the PSA program. I do want to say that earlier, last year when this came out, it was important to me that we give this out to the region. So fortunately, working with Chuck, we were able to procure– I think in '17, I would like to address the procurement process. For the first time ever, I had to bring cash to give to Chuck. But he promised me that I got a discount. I think we only paid $500 a copy. [laughter]

But our regions do have this. And I will ensure that the regions get this. And it's an area that we're going to continue to support, and [12:16] the ongoing relationship. [applause]

Panel: Recent State Legislation and Actions

Mr. Ron Jacobus representing Sen. Bob Hall, Texas
Hon. Andrea Boland, Maine
Harris County Texas Emergency Management Commissioner Robert (Bob) Janusaitis

CHUCK MANTO: I just wanted to mention a few words about Senator Bob Hall and Ron Jacobus. They've been working on these issues for some time in the state of Texas. They had a Secure the Grid conference just a year ago, right?

By the way, any of you come up here, we're missing our initial step. And once you get here, the chairs have to be arranged so they don't fall through the cracks in the risers. We've had people topple in the middle. It's part of a comedy routine to give you some relief from the seriousness of these discussions.

But anyway, Senator Bob Hall convened at the state capitol a review and assessment of the risk facing our critical infrastructure. You know Texas, because they have their own grid that's separate from the rest of the country. You mentioned that as well.

But there's a lot going on in Texas. Ron can talk a little bit about what's about to happen, and then, similarly, Andrea Boland got things kicked off at the state level in the state of Maine, and she may want to give us a couple minutes' overview as well as to what is and what isn't happening in the state of Maine; and, any other place that you may wish to discuss. And I will be here to sort of help answer any questions you might have.

I'd like to try turning it over to Ron, and you can either speak from there or here, whatever one, whatever you feel more comfortable.

RON JACOBUS: All right, I'll stay seated. Thank you, Chuck, for the invitation to come here this year. It's an honor for us to be here today. Nik Hanlon, can I have you stand up real quick? Nik is probably one of the most influential reasons why I'm here today. So he took me on as an intern in 2014, and I had nothing but a college background. So I got into EMP with him, and then Center for Security Policy brought the Senator and I together in Las Vegas while [2:03] and working on EMP ever since.

So I'm going to try not to make it sound like an Emmy Award ceremony, but I know a lot of people in the room here that have helped us out and brought us to where we are now, a place of confidence, as we move into the legislative session starting next month. And we are bringing EMP legislation that covers not just the electric grid, but also our emergency management at the state level and the county level, down to the local level. So that's why I'm happy to have Bob Janusaitis here who's worked with our office pretty much almost from day one, since I moved down there, to help the Senator on this issue.

So I'd like to also thank Director Breor, and actually highlight Scott and Michael, his PSAs down in Texas. And we've had a chance to meet with them, work with them on this issue and others, and they have been a great resource. So for anybody who's not using your PSAs in your state, please do. They're there and they're willing to help. And we need it from a legislative perspective.

In addition, Chuck Manto and InfraGard have been probably the best credibility bump that we could get as an organization in partnering with the FBI as a public/private partnership. And for somebody who has a formal designation to take [3:28] once you start bringing up an issue, to have a bipartisan/nonpartisan organization that could just stick on the issue and talk to everybody, they don't have the same limitations that we do as an office and the hurdles that we encounter. So it is a great resource. So for anybody that's watching this and in here that is an elected official, please use this as a resource to get this issue out and others.

Peter Pry, who was with the Senator prior to my time with the Senator and has been a great briefing resource.

And in addition to that, Tom Popik, who I will not steal his thunder, he's been another great resource in the Resilient Societies in helping us with the cost estimate, as we know cost will always be an issue and always be a discussion that needs to be had as we move forward with this item. And so he's kind of jumped out on a tree branch of his own to get ahead of the curve here. Because everybody wants to talk about the cost and no one's willing to do the cost model. So his organization has stepped up and at no charge have helped us out in a place where we wouldn't have been able to do it on our own.

And George Baker has been an instrumental part in that study as well. Also, Neil [?] is in the back there, I can see him. And Brian has helped us a lot with briefing on this issue. And Mike Carusso as well, another EMP expert, has the hardening and technical expertise that I lack. And they're able to crunch the numbers and explain that to us, and then I can get it into layman's terms for everybody else.

Brent Collins and Ben's organization, they have helped us out tremendously as business executives in national security in getting that messaging outside of the technical world and into the kind of leadership discussions that need to be had in the executive level [5:30] business-minded folks back into the security setting on protecting your infrastructure so that we don't only protect the investments that we have, but also lives that depend on it. When we talk about EMP as a threat to the electric grid, it's always tough to get some of the business-minded folks into the life-sustaining and lifeline infrastructure mindset.

I'm going to stop there for a moment and give everybody else on the panel some talking space. But I'll jump back. Just wanted to thank those individuals that got us to the point where we feel confident in our legislative area as we move forward to the subject.

ANDREA BOLAND: This is just a brief overview, and I don't have my notes anyway with me. But my name's Andrea Boland, and I'm a former state representative from Maine. As many of you may know, I had the pleasure of sponsoring the first piece of legislation to get enacted in the United States on EMP and GMD protections. And I could just recite the same list of folks that Ron Jacobus just recited as being wonderful supports in Maine. And others – Cynthia Ayers is here also today. Bron [?]– I mean, Bron Cikotus was also wonderful, and unfortunately he's no longer with us. And another wonderful science resource that I have that has helped me with a lot of things, he just goes by the name of Kirk.

So it was very exciting to get that passed. That was in 2013. And it was considered landmark legislation, long legislation, and these folks are the ones who really helped deliver it. I brought it, and I shepherded it, but there's tremendous expertise in this room to bring it forward.

As some of you may know, I recently failed in my attempt to unseat the Senator who represents my district this time, and who, interestingly enough, was the one who actually blocked Maine's progress on getting EMP protections on the grid. He was, and likely will continue to be, the Senate chair of the Utilities Committee. We thought we had something that was pretty much going because he had been saying all along he was supportive of it, and voted for it in committee. I told him his leadership was going to give him some difficulty. In this case, that was Republican leadership on the Democrats. So it's not just Republicans that are leading on this thing. And he said he could stand up. And then in the end, at the very end he stood up in his chair of the utilities committee, gave a speech saying, We ought to kill the bill, it was just too expensive.

This was not my bill; this was last year's follow-on, which asked for specific protections, the neutral ground blockers, which were shown to be effective. GIC monitors, which we didn't have. And something else I'm not remembering at the moment. But things that came from studies that were done pursuant to the legislation I brought that had been identified and at the last minute passed the House fine. Got to the Senate and it lost by one vote. And so, of course I was unhappy with that guy.

But late in the game, I was asked to go in and run against him, and I was happy to. But I did not succeed. It was not a good district for me.

Since then, the project director of the ongoing study team that had persisted, of which Primus and resilient societies and I have been participants, the project director, there was a second project director who left. Both of them were really good. And I'm afraid since that last project director left that things have slowed down even more.

Also, the governor's state energy director recently announced his retirement, and I quote from the *Portland Press Herald,* expressing frustration with the power of lobbyists to define and advance energy policy in a citizen legislature. He's right. The problem that we really have in front of us is regulatory capture. And it's not just at the state level, it's at the federal level as well.

So we have made some noteworthy progress. In spite of some of these halfwits that we've had, the legislation that we brought resulted in studies, and the studies have been ongoing. And we do have some improvements on the grid that help the energy flow more smoothly with series capacitors and reactive power support. But CMP never did do half of the study which had Central Maine Power, did never do the half of the study which was to look at EMP. It only was willing to do solar power, solar events. And the public utilities commission, the one who was most in charge, supported them, just went along with that.

So now we have a new project director, and what he reports to me in a conversation, recent conversation is that the GIC monitors that were put on were seen as pilot issues have now been decided that they were going to be permanent, and they're going to have them now throughout the whole system for Maine. And that's progress.

They've also just signed a continuation of the agreement they had before with an outfit called Pingpings[?], which has been responsible for analyzing the data generated by our new synchrophaser unit, which gives a picture of the transformer performance during GIS flows and harmonic stirring solar storms. This is real progress. The data that's generated is enormous and Pingpings has done a great job. They've also applied for a grant to enhance their equipment and Central Maine Power is a partner with that.

So those are the positive things that I see. However, Central Maine Power is also releasing their finalized report very shortly on all that has been done. And this has been two-plus years of work, including outside experts that were specifically requested to be part of this study team by the utilities committee. And they find no concerns with their system. Because they are going by NERC standards. And it was the problem of NERC standards that led to the issues that we have right now throughout the country. They're satisfied that the model they're using provides a good representation of the activity based on the lower level storms that they've experienced. I asked if they did any higher level testing, and actually they have, up to twice the level of the NERC standards, and they propose to do even higher. So that is something.

They just say that their big thing is to meet NERC standards, and they are looking at going beyond that in their testing, and they have done that. But they are slow-walking it, as Peter would say in his experience. They have no immediate plan for the neutral ground blocker pilot that they had talked about before. But they are going to investigate what it would add to the system.

So it's just a slower process. At least we still have the neutral ground blockers in the equation and they are still looking at it. And I think that that's helpful.

But the report that's coming out, they've said to me, does not include anything except their own work. The study team was supposed to include outside experts. And Primus did fabulous work for the state of Maine and produced its own report with support from resilient societies, and apparently that's not going to be part of the final report. I'm sure there's going to be a lot of outrage seen by certain legislators and other experts as a result of that.

But to conclude, CMP continues to work with Pingpings, which is great. It's adding more GIC monitoring throughout the system. It continues to be out of conformance with doing EMD as well as GMD studies. And of course, they're out of conformance with the intent of the legislature to use outside experts in the team. And their expertise has been enormous and highly, highly valuable, as Ron Jacobus has also said. And they are adhering to the NERC standard for their own benchmarks, even in light of what I would call its treasonous understatement of the threats.

We are essentially in a war with nationwide regulatory capture because even though, as I saw in Maine at our hearings, the utility companies have little interest in doing this. They have little expertise. They're extremely embarrassed. The utility committee was shocked at what they didn't know. And yet, still, public utilities commissions and people in government turn to them as their experts. They've got to include the outside extras who do their work independently of any other concerns.

So thank you very much for your attention. And I congratulate the hard work that's going on in Texas, and I continue to work with private businesses and other authorities in Maine. [applause]

RON JACOBUS: So I'd like to pass it to Bob, who is leading right now an effort to put together a tabletop exercise in Texas. Unfortunately, the tabletop exercise is a triple threat workbook, although genius in its thought, already assumed that Texas was protected and we had already done the work we needed to do. So we had to go back and have Chuck and Mary give us a new scenario where Texas is indeed down and we have enough protective system for our emergency managers, Bob being one of them, to go through and work this through at state, county and local levels. I'm going to let him talk about the setup work he's doing currently to help us out with this issue in Texas.

BOB JANUSAITIS: Thank you, Ron. Just to set the stage, the Texas legislature is only in session once every two years for 180 days. So they can [:55] But also, it means that there's a lot of work that has to happen in that session over 180 days. Senator Bob Hall has already prefiled a bill for protecting the grid and critical infrastructures.

What we're doing, legislation was attempted to go through last time, it went through the process very well, but it didn't get completed. This time we got a much quicker start on it But what we're doing is we're putting together additional information and leveraging some of the resources that Ron talked about earlier.
But we're also going to be doing a tabletop exercise in mid-February. We couldn't do it any sooner because of the Super Bowl and we're occupied by the Super Bowl. But in mid-February we're going to be bringing together approximately 30 invitation-only players for this exercise. It'll be various counties in southeast Texas where we're inviting state of Texas, DHS, FBI, [1:53], a number of others, Texas Medical Center to this exercise.

The intent of that is to just throw it out on the table, look at what the potential consequences are and rapidly turn around an after-action report that we can present to the legislature at about the end of February, which will be right about in the middle of the legislative session, with the hope of getting this out there and attention to them so that they'll actually see it and say, Wow, this it's not only the [2:21] out there in other parts of the country, but we actually have expertise here in our state that are looking at this thing, and it's a real threat. So we're hoping that that will kind of get us over the hump and this legislation will be law about next June.

RON JACOBUS: The idea with the tabletop exercise is to serve as a catalyst to a provision in our bill and it's focused on Texas emergency management. And to standup a task force within TDEM that will be comprised of ten individuals, Bob Helber[?] and others, and talk to outside experts. We've heard from the EMP Commission members an awful lot and we'd like to get their input for the Texas perspective. But from [3:08] preparedness, what are we going to do if we're an unprotected system, if we're in a partially protected system, [3:17] And how are we going to mitigate the disasters? If you come from EMP [3:23] take out a grid for 30 days or more.

So that's really our focus. Use our resources within the state and then outside of Texas as well and leverage that to the best of our ability to get the legislation passed that we need to start working on protecting the grid of Texas.

For those who don't know, Texas is unique for many reasons, but one of them being we have our own self-contained electric. It's called ERCOT, the Electric Reliability Corporation of Texas. So we operate our own grid that can actually be cut off in a disaster scenario, really any scenario that Texas wants to shut off its power and limit it to the [4:11] We have that ability. We also have the ability to detect without relying on anyone else, federal or state, anything that would serve as a hurdle in addition to the hurdles we already have in the state.

So we're pretty excited about the opportunities we have, and as we move through the legislative session, I think we've done a really good job on educating on it. The Senator likes to say, We brought this issue to bear last session. There was nobody else in the legislature that could spell EMP, let alone know what it meant. [laughter]

And I think things have changed. He gets stopped when he's outside the legislature in town by individuals that ask how are we coming along on EMP legislation. He can stop and [4:55] legislature now in the interim [4:57] members how is the legislation coming and we prefiled it, everybody's starting to take a look at that. So we now move to the [5:06] meetings with the ERCOT folks, the ACU, the POC, and see where they're willing to make some changes. Are they willing to bend [5:17] what our line in the sand is [5:20] on this area here. We're going to protect [5:24] Texas and [5:26] the model for the rest of the nation.

With that, you cannot have reached where we're at in the education area without the help of InfraGard, without the EMP SIG that operates under Bob [5:43] in briefing city council members and county commissioners, all the way up to elected officials at the state level as well. So our education's been a key hurdle, I think, not just in Texas but everywhere else. And that's always tough [6:02] once you do that [6:05] implementation that follows.

We have many of the solutions figured out. It's just selling that to legislators that are kind of in a new position of maybe passing something that's more than an insurance policy, that's down the road when they have immediate concerns that aren't necessary [6:24] look at them from the EMP [6:26] whole society [6:29] in their state as well. [6:35]

Chuck's ready to roll along here, but one last note.

CHUCK MANTO: We've got a couple of questions [6:40]

BOB JANUSAITIS: All right, then I'll save my comment for asking questions.

CHUCK MANTO: First of all, I want to give a special word of appreciation for the entire EMP SIG for what you folks are doing as leaders in your own states. Can we give them a round of applause? [applause]

And the reason that this is so important is most of these people, when you look at their job descriptions, don't have this in their job description. And the problem is, nobody in the country does. And so when your realize that there's a huge problem and there's a gap and it's nobody's job to get it done, you just sort of let the country die off or do you let the community fall apart? Or do you step up say, I'll take responsibility for this. And this is exactly what they've been doing. And Texas, of course, is notoriously independent state. In fact, they're not really a state, are you? What do you call it? A republic.

So it dawned on me, I was over at the state capitol in Texas and we were doing the Pledge of Allegiance. What's the first phrase: I pledge allegiance to the flag of the United States of America. And to the republic– now, I figure a Texan wrote that. [laughter] Threw that phase in so we have to be remembering Texas every time we do the Pledge of Allegiance. But I assume that wasn't the republic they were referring to, but maybe Texas can elucidate.

RON JACOBUS: Well, you recall we didn't have a pledge to the Texas flag right after.

CHUCK MANTO: So you do a lot of [8:15] okay, great. But anyway, I just want to show special appreciation for Texas. And I also wanted to mention a few other things. You'll see from what they said, those of you here rallied support for like Remember the Alamo, and everything else. [8:29] And in fact, Texas, it is now intended, and this has been helpful to folks elsewhere as well. And we've had folks from here helping the state of Virginia. We've had folks from this room helping the state of Maine. We're all in this together. And most of the people who have been helping you have done it with no pay. And it's not because they were bored, with nothing to do; they were putting their lives on the line

Again, how many former military in the room? So what people don't often realize is that you're willing to put your life down on the line to get something done. Very few people are not willing to do that. If I were to be a Fortune 1000 CEO and I lose $100 billion from my company, am I supposed to take personal responsibility? Does that mean I write a check to cover that 100 billion? No. Does it mean I cash in my house and go home with a street cart to take care of my shareholders or my employees? Do I let them get fired first before I even get my bonus cut. Get all of you who served the military, who were not only asked to do something like put your pension on the line and your future earnings on the line, you're asked to be willing to die.

Now, those of us in the EMP SIG who have been willing to help each other, and you have helped us sort of model that willingness to put your life on the line, pay the price of these other people's needs. And it's very applaudable, and I want to personally thank you again for doing that.

You had a comment that you wanted to say, too.

RON JACOBUS: Thank you, Chuck. I'd like to flip that and put it right back on InfraGard, and especially the Houston chapter with InfraGard. You have some real rock stars. Bob Janusaitis is one of them. Michael Lambert who operates within the EMP SIG as well. But really, I have to hand it to Angela Hahn, the InfraGard coordinator and sponsor for Houston. She is an absolute rock star. And she takes us, it's not just [10:28] but she really has put forward some really [10:31] great thinking ideas to allow us to come over there and use it as a resource.

So we could not be where we're at right now as an office in the state legislature without the [10:46] So I'd really like to throw that out and say thank you.

CHUCK MANTO: Thank you very much. [applause]

RON JACOBUS: And since this is the 20th anniversary of InfraGard, I went back and checked what I was personally jazzed about in 1996. And it was *101 Damnations*, the Disney movie that had come out back then. And *Mighty Ducks 3*. So just put in perspective where I was when you guys were starting this 20 years ago. We thank you for everything you guys do.

[applause]

Report from the New EMP Commission

Dr. George Baker

CHUCK MANTO: –Dr. George Baker has an illustrious career, lifelong, learning about these phenomena, working for decades on this issue through the Department of Defense. He also after retirement continued to work as a professor at James Madison University out of Harrisonburg. He's been a substantial guiding person in his own community in Harrisonburg, who has an exceptional amount of resilience themselves. Dr. Baker, come on up and feel free to give us an update on whatever you think is most appropriate.

GEORGE BAKER: When Chuck asked me to say a few words about the EMP Commission, because we really haven't been recognized, the new commission has still not met the budget, we only recently realized the budget. I just got an email where I still have to sign a bunch of papers, nondisclosure papers in order to be on the staff [1:20]

So I thought it might be better for me to review the findings of the last stage of the EMP Commission. The Commission met for 2002 to 2008, it made a number of recommendations. And I just thought that might be instructive, to review what the original recommendations are maybe assess progress, or lack thereof, in the last eight years since the Commission closed its doors.

This is the original EMP Commission. Bill Graham was the chair. And the same members are still alive and have agreed to be part of a new iteration. Their conclusions [2:27] EMP is one of a few potentially catastrophic threats to the United States. The end of the world, as we know it, what's the acronym, EOTWAWKI, end of the world as we know it [2:39]

By taking action, EMP consequences can be reduced to a manageable level. And the US strategy to address EMP threat are balance, prevention, preparation and recovery.

Now, *[pointing to slide projected on screen]* I have a little code. If you'll notice, each one of those main bullets, there's a symbol. There's either a tilde or an X. Anybody want to care to guess what those mean? A little bit and nothing. So first recommendation, these all support the three major bullets we just talked about. Pursue intelligence interdiction and deterrence to discourage a nuclear attack. These work in CIA and within the Department of Defense, Pentagon and [3:38]

And highest priority, obviously, is preventive factors [3:51] Because this is so extremely isometric, you only need maybe to take out one or two [3:59] There has been a lot of, as you saw, a news bulletin this morning that there's a new multitarget system that's being developed [4:18] It looks very promising. Ballistic missile defense also [4:25] some major progress.

Shape the global environment to reduce incentives creating EMP weapons. I say we've pretty much [4:35]

Make it difficult and dangerous to [4:37] I don't think so. I don't think [4:40]

And next one is protect components of key infrastructures. That's not happening. There are some utilities that taken [4:53] And I can count those on one hand where we've actually protected some data centers. But besides that, the large majority, probably 99%, is unprotected.

Maintain the ability to monitor and evaluate the condition of critical infrastructures. [5:12] surveillance. Well, since we haven't protected much of anything, there's really nothing to monitor. So there isn't [5:17]

Recognize EMP attack and understand how it affects differently from other disruptions. I've got to say I give a lot of credit to InfraGard and some of the watchdog groups out there, foundation groups, resilient societies, EIS. And there are others.

Plan to carry out systematic recovery of key infrastructures. Again, for EMP that's not happening.

Train and evaluate [6:01] and report to Congress. That is happening within DoD. DoD, I've seen these reports, I still have access to them. DoD has an annual report on their progress. But I have not gotten [6:18] Over the last year or so DHS [6:26] DoE's got a program, also. [6:32] ramping up. So within the last year, I'd say that the other agencies are getting on board.

Define the federal government's responsibility and authority to act. The federal government is still in an organizational mode. That is not happening. The EMP Commission specifically mentioned DHS and DoD responsibilities there. As a result, a lot of the work is devolved into the states. And I'm very pleased to [7:16]

Recognize opportunities for shared benefits to protect [7:22] That hasn't happened. What we have is still [7:31] people that are looking at GMD without bringing EMP into the mix, and that is counterproductive. GMD protection will not protect you against EMP. EMP protection will handle GMD effects, but not [7:53] And so, with these stovepipe efforts, we're setting ourselves up for spending a lot of money and really having an unprotected infrastructure. We need to [8:05]

Conduct research to better understand infrastructure system effects. Yeah, that's happening. There is some work at [8:19] national labs. The Department of Defense has a lot of research work going. Now it looks like DHS may be doing some research. [8:33]

And then finally, the Homeland Security Council was asked to prioritize the government society-wide efforts to counter [8:48] threats. I'm not aware that that is happening. And if that's happened, I have seen nothing [8:56]

Department of Homeland Security established a dedicated program. I have to confess, this chart [9:07] So there is some work that's happening at DHS in the last year. They have a specific program. I'm not sure it's a dedicated program. Maybe they will be hard on them.

Establish senior leadership position, accountability and authority, appropriate resources. I'm not sure that's happened. Metrics were set [9:33] Hopefully that last X [9:43]

Finally, the project has been slow. We accept the EMP Commission recommendations [9:58] There's much more to be accomplished. There are large impediments that exist. There's prevalent misconceptions and misrepresentation of the effects. This idea of EMP effect as being unbounded, I don't understand that term. It's well bounded.

Stakeholder reluctance. We have industry reluctance to protect. And there's no one in charge. Again, we have all these sort of amorphous, different things going on at DHS, DoD, DoE. But there's no cohesion to what's happening.

So the EMP Commission has its work cut out. That's why they're still at this stage here. And I think one of the most important things we can do is what we did when we were [10:52] in 2002 to 2008, is to provide a monitoring and review. It's amazing, when agencies know somebody's watching, how they may progress.

[applause]

"Powering Through" and the Dec 1 Workshop

Mary Lasky, Bill Harris, Steve Volandt, Fred Ferrar, Ron Jacobus

CHUCK MANTO: So now, if our next panel can come up. Actually, Mary Lasky, who's been working over her career at the Johns Hopkins applied physics lab in various roles in IT and business continuity. Their office supported us when we were first initiated in 2011. And we did a tabletop exercise and workshop with the National Defense University at Fort McNair in Washington, where we looked at a high impact case and a low impact case of EMP. We did the mid-level EMP case at Johns Hopkins applied physics lab, and Mary was instrumental in organizing that effort for us.

What I'd like to do is have her come up and introduce her panel. She's been working tirelessly for the last couple of years, working with us on a weekly phone call, in addition to our organizing call we do on Friday, they have a Thursday call that works on these publications. And this year she's spent a substantial amount of work to be able to provide sort of a breakthrough book.

I'd like her to discuss what that book is, and then have the others who are participating with her on this in the workshop that we did yesterday at the National Guard Association of the US headquarters about our next steps and how you might be able to participate with them.

MARY LASKY: Thank you, Chuck. Could I have Steve Volandt and Ron Jacobus? I don't see them, but– Brad and Bill, would you please come up? We'll take it in the order that we get them.

I would like to first start this off by saying something about Y2K. I heard this yesterday, and I heard it again today, that, oh, it was wonderful. And then December 31st happened, and then we rolled into January 1, 2000, that nothing bad happened. And it was known as Y2K. That "nothing bad happened" because we as a nation took a huge amount of effort, a huge amount of publicity about what could happen, and we hired lots of people to go and work the situation. So we as a country thought it was very important. And wouldn't it be wonderful if we as a nation take this issue of an EMP and take it as being so serious that if it happened, nothing would go on the next day. Wouldn't that be a great goal to have for this? So I think when they mention Y2K, we have to realize how much work really went in to it.

So let me go to 2015. What happened in 2015 was that we had a workshop and we kicked off some ideas about how to be protected if there was a long-term outage of the grid. So we looked at long-term outages being weeks, months, or even a year. And we said that's widespread. So that it meant that it wasn't just localized, but it could be all the nation.

So we said that this was really important, and we started working on this. And you can see that we presented the book, *Powering Through.* You can get this out at the front desk. And we are really very excited that we've had such great endorsements, one of the endorsements from Ted Koppel, who had just said how bad things would be in cyber, says, "*Powering Through* is a gigantic step in the right direction." So we're very excited about that.

This book wouldn't have happened except that we had 26 authors. And we had a great number of people who were on the weekly phone calls, conference calls. And what we did in writing the book was that somebody would write a section. It was sent out to the list of people. And then that week we would have, or the next week we would have a review of that. So each author stood a review of many people on those conference calls, talking about what they'd done. And we would refine it and have new versions of it. So that was exciting.

I want to say thank you to Chuck Manto, because he gave us the confidence of letting us go off and do this, basically on our own, without a lot of his involvement. He would check in every once in a while to be sure that we hadn't gone completely rogue. So a big thank-you to Chuck for your support and your encouragement on all of this.

I would like to thank also Steve Volandt and Bill Harris, because they have been– the book wouldn't have been possible without our work.

The book discusses the grid and what happens with the grid. It also discusses some of the, as Scott Breor said, some of the documentation nationally. And then it goes into threats. And we talk about five different threats. We talk about EMP, solar storm, we talk about cyber, physical attacks and then regular frequency weapons. So all the five.

And in the front of the book, in red, is the consequences of those five different threats. And this is the George Baker chart. And all of that is explained in the book. And we look at severe kind of threats to the grid, and then not so severe threats to the grid. But if we're prepared for the severe, we will be prepared for whatever comes along.

The book looks at interdependencies of what we're going to look at, and it looks at some of the critical infrastructure. One of the things that we found along the way was, water was so important that water had to be really critical. If we keep our water going, then in our cities we can keep people at home, and we can keep our waste water running also. It's extremely important. So it's very definitely talked about in the book.

Another important, I think, new thrust in the book was resilient community islands. And a resilient community island can be your community, it could be a university or college campus, it could be a hospital complex, it could be a business park, a faith organization. So we came up with many different ways that we can have resilient community islands.

Chapter Eight in the book actually talks about what if this were to happen, what would we do? We look at whole communities, starting with citizens up through the government. Looking at cities and rural, looking at businesses, looking at nonprofit organizations; so really, whole community.

So we published the book. Here it is. And that's exciting. And as we started thinking about it, it was a situation where we would say, oh, my goodness, we wish we had had more in this topic, or more in that topic. And so, we said, well, this is just volume one; we've got to move to volume two. So yesterday, we had a workshop on this. And so, I would like to introduce the team of people. And I don't know where Ron Jacobus is. Bob, can you go find him?

Bill Harris led a team looking at alternative resilient strategies. Steve Volandt looked at resilient community islands. Ron Jacobus looked at the government focus. And Fred Ferrer looked at critical infrastructure. So would each one of you– Bill, would you start and talk about some of the results of what happened in your breakout session?

BILL HARRIS: Thank you, Mary. So we looked at the types of modeling, the methods that could be used to improve our understanding of the interrelationships among these critical infrastructures, the consequence management choices, and how that feeds back into priorities for investing to protect our infrastructure.

We looked at the conventional system, is to have reliability standards for any particular infrastructure supplemented by best practices. Even there, there's some important infrastructures that don't have reliability standards, such as the interstate gas pipeline. No one's responsible for reliability, so you can put in an all-electric compressor, which will compound the power outage, as we found in Texas in 2011, and in California and Arizona in 2014. So we know that's insufficient.

Then we looked at modeling interrelationships that are in pairs of critical infrastructure, the water/electric interface. But even for agriculture, if you look at a pair it doesn't work because you have to look at water, electric and chemicals. Then you have to transport it. So we were especially impressed by the modeling opportunities that [10:35] is doing and that others of us can participate in, to do model-based system engineering, which can begin to put first a document, what the infrastructures are now, what are some of the relationships, including invisible, critical interdependencies. And the synergies – as you fix one, what are the benefits for another?

So we don't think there's a preferred method. We think all the different techniques that we've utilized are worth doing. But we did come up with some conclusions yesterday afternoon about where we think Version 2.0 of the book should concentrate for improvements.

We wanted to improve what we knew about the electric grid and its relationship to other energy sectors. There's been work at the EIS Council, two books by Paul Stockman that looked at fuel and water. So the energy sector.

We are weak on communications in this version. We know that many communication systems being out will cause deterioration of all the other resiliencies and recovery opportunities. That's a second area.

We still want to understand better the water and waste water systems. And transportation as well. But we then found that there were other critical needs that warrant an entire infrastructure. The positioning and timing systems from the GPS system are vital to basically all the critical infrastructures. And we probably need a land-based backup with GPS system, or some combination of that and better EMP-protected– it's more the ground stations than the satellites. Some of the satellites already have some protection, but the ground stations are vulnerable.

So we thought that was important. And then we also recognized that there was a special role for the National Guard. The National Guard already has taken a lead in cyberprotection. They're being increasingly trained to come in to protect against cyberattack. But there are so many other sectors where they could make a big difference.

We talked briefly about the food distribution system. It will be vital– I forgot to mention the agricultural section. We're especially weak on agriculture and the food supply chain. We found in Version 1, and we were surprised to find that we had gone from a system that not only had MRE rations for military personnel, but we had a system of rations in public storage for civilians. We've had it from before World War II, but we had it especially 2004 to 2008 with concerns about anthrax attack. We have warehouses that will provide a real cushion in the event the grid was lost.

The EMP Commission reported in April 2008 that we needed to have stocks of food supplies because we would lose at least one growing season after an EMP attack, or maybe more. And our ports have become, they used to have diesel fuel; almost all the cranes are now all-electric. Again, we have several situations where air quality decisions drive us to greater vulnerability, reports on the all-electric compressors on gas pipelines and others.

So we found that in 2008 we went from a plan to a framework for emergency agricultural supplies. A plan you have to implement. A framework you can pretty much do what you want.

So that last year of the Bush administration, we had that change. And in 2009 and '10, the first two years of the Obama administration, we had moved to 15 regional supply systems for emergency food stocks, and those food stocks were distributed and not replaced.

So we're down to about ten days in the supermarket supply chain system, and another ten days, at most two weeks for agriculture. Which means we don't have– we can't fulfill the goals to have shelter in place in large urban areas that have over half our population. So that was another sector. If we can generate some money from the book and possibly from other financial grants or partnerships, we want to have some workshops. We want to go back to California in the food supply systems.

We know do not have much for financial system resilience, but we did reach a tentative conclusion that we might be able to utilize the next generation of smart phones as a basic building to recover financial transactions from contracting. The older smart phones that were available in 2007 were EMP tested by the old EMP Commission. And if they weren't plugged into wall outlets, they almost all worked. But the cell tower, the entire communication system would burn out in an EMP attack.

So one possibility is to work with Silicon Valley designers on the next generation of smart phones. Make the phones a little better EMP protected. But also then work with the telecommunications industry to identify places– Ross Merlin yesterday suggested the Postal Service to have a place in each community. Everyone knows where their post office is, or to have the banks in a community have at least one place where you could bring your phone and be able to communicate with protected networks through the data centers that are protected. So that's not going to happen immediately.

But if we could recover the financial system, so we are not into a barter economy, it should immensely improve the rate of recovery. It's just a much more efficient economy if we have the market allocating scarce resources. And you can't do that without financial transactions.

Then the last things that we considered important were, I had mentioned the National Guard, but more generally having the Department of Defense take some responsibility for providing standards and support for the civilian infrastructures. Back in 1966, after the Cuban missile crisis, Castro would not agree to limit missiles, and there was an issue of whether utilities in Florida should have the responsibility to protect against missile attacks on electric utilities, especially the nuclear power plants.

The predecessor to the Nuclear Regulatory Commission adopted an enemy of the state doctrine that basically said that if there's a state threat, that is not the responsibility of the utilities to defend against it. That has the perverse effect of making it more expensive if you don't design protection of your control center; the cost now soars from 3-5% to maybe 20%. So we need to bring the Defense Department back.

So for our book, we find basically some priorities that we want to work on. Thank you.

STEVE VOLANDT: So our breakout session involved resilient communities. The benefit, when you think about this problem from a community perspective– from outside a community perspective, it's overwhelming and complicated before you start. If you put it in terms of a community, the community becomes the focal point of the infrastructures and the points of sustainability and resilience that come together to support that exact community.

So that's a very generic idea. What is a community? Every community is different. So how can we standardize this conversation in a way so that we can compare our unique communities, or our unique organizations which can be looked at as a community, against a standard set of ideas that help you instantiate sustainability and resilience.

The difference between sustainability and resilience is sustainability is thinking about how much food you have, how water do you have? Do you have spare parts? Do you have people coming to work? Do you have fuel? Resilience is protection-oriented; it's in the security aspects. So are your systems EMP protected? Are they cyber protected? Are they protected against physical attack?

Those are the aspects that we snap into the community concept. In our breakout session, we talked about coming up with some standardized models of different community types just to use as examples in order to be accelerators for planners or for people to get their head around this idea. So examples of a sustainable community model or a template that we would document could include hospitals, perhaps a small, rural town, a university campus, military bases, and so on.

There's a spectrum of community that ranges from, on one end, a very rural, perhaps even Amish, who don't need electricity and they do everything in-house; and then the far end of that spectrum would be a manufacturing facility inside a very large metropolitan area that is reliant on the global supply chains. So the two of these can run the entire spectrum. It's easier to model the simpler one [21:25] So it's a good place to start.

We can look at each of these community sets as almost like Lego blocks. In a way, our current infrastructure is all interrelated, interconnected. And it's not segmented. And therein lies our dilemma. We can take our infrastructure down from single points of failure. If we're segmented, similarly to inside a ship with watertight compartments, you can seal off the [21:56] zone, and other parts just don't work.

So the concept is: let's get some resilient communities established. They don't have to be built from scratch. We can overlay and retrofit resilience and sustainability into some communities that we have today. Just gaining an understanding of what would be good starting points. Would it be a city block? Would it be a rural agricultural community? Would it be the zones that surround critical manufacturing?

Identify how sustainable these communities are. Can they last a year without resupply? They can't. So how far can they last? Take inventory. There's a process to this. Take inventory of what they're capable of today. Compare that to their ability to be a resilient island or an island that requires interconnections with other islands of resilience. Understand those gaps in sustainability and resilience and let's go ahead and close those gaps.

I'm keeping it brief, but we need to recreate the buffer that used to exist in the United States when we had lots of family farms and lots of local production. We've become adjusted to a global economy, and that has become our Achilles heel. We can resegment and create more local communities that are microgridded and can power themselves and produce more power and food and water [23:32] And get enough of those clustered around dense metropolitan areas, you've got a great start.

FRED FERRER: I'm up next. I don't have any time, so that concludes my presentation. [laughter] Mary, I'm going to go ahead and do what you just said, two to three minutes, max. Myself and Steve Pappas led the largest group that met yesterday. Steve's in the back; hey, Steve. And who out here was in the workshop? We had about 30 to 40 folks.

Looking at the private sector, we all know all of this is meaningless. We can talk about professional activity, we can talk about DHS, FBI. We need the private sector, the owners and the operators of all of this critical infrastructure, right? So a lot of our discussion yesterday talked about what we need to do for Version 2, is to involve the experts from the private sector who can really flesh out what it is we need to do and what we need to say.

Critical to that is, what is our message? What is our message. And in developing a communication strategy, who are we talking to? Who are we trying to get involved? Who are we trying to educate? So we need to think through those things.

We went from communication and audience to talking about some of those people who we like to involve in Version 2.0, and that would be using some of our existing resources in InfraGard, such as the sector chiefs, to pull those experts in closer to us. Also, in my opinion, and in the group's opinion, we need to be working much more closely with DHS and PSAs. If we don't know who our PSAs are, if we're not tied to the hip with them, we're failing in what we need to do. Because they're the ones that are really committed to pushing this forward.

So involvement with the DHS, PSAs, sector chiefs. Trade associations, not-for-profits. And one of the big ones, and what we, I hope, we really started to flesh out here in Version 2, is faith-based organizations and communities, some of those like the Mormon Church that has a lot of preparedness and resilience and security. The Red Cross, a not-for-profit. So we need to involve those faith-based organizations as part of this resilient solution, this puzzle, maybe the creation of the islands as we move forward.

And that is it. Two minutes.

RON JACOBUS: I've got the government domain, so I don't know if I can do a two- or three-minute windup. [laughter] We had a very spirited and lively debate during our breakout session. And I would like to liken it to trying to put a man on the moon in about an hour-and-a-half. The government domain, I would say, with respect to the other breakout sessions, may be perhaps the most difficult domain area to flesh out. As Roscoe Bartlett and Peter Pry mentioned, they brought this forth to Congress back in 1997. And today we've got no federal legislation. So we need to start looking at what we're doing, how we've done it in the past. Is it working? Has it worked? Do we need to change some of the strategies? What strategies do those need to be?

To use a Texas model, you can drag a horse to water, but you can't make it drink. People are going to do what they only feel they're incentivized to do. So when we talk about the government domain, we look at elected officials, getting nonprofits to act as a buffer between constituencies and the elected officials, to educate the elected officials and the public to say this is something that's easily done. And once you have a general public that is staunch on the issue as a whole, you tend to see government acting in response.

And so, getting government to move on something takes everyone in government, especially in this scenario of [27:46] threats and protection from state and local to federal. In order to row the boat, they all have to be rowing the oars in the same direction.

So that's kind of what we got to. And we need to figure out exactly how to do that in 2.0, and how to get over the hurdles that we have had in the past.

So thank you.

MARY LASKY: Thank you. And thank you, each one of you, very much. [applause]

Implications for Best Practices

Chuck Manto (with *Powering Through* Panel)

CHUCK MANTO: What I'd like to do as part of my time, talking about implication for best practice, is have one or two microphones in case there are any questions that you'd like to ask them. The reason I'm saying it is, we'd like to have your involvement over the course of the next year on these next publications. You're welcome to network with them informally during the course of the day. But I would like to make certain that if anybody had a question or two from the audience that you'd like to raise to them, make sure you identify yourself into the microphone and use the microphone.

ED HAMMOND[?]: I'm Ed Hammond from Center for Security Policy. My question is for Bill Harris, and anybody else who'd like to add. I guess it's a follow-up to the government or the messaging and PR concerns. Can you guys take a stab at just describing the difference between risk analysis and threat analysis and the cross-cultural bridge that has to go from military to the private sector.

CHUCK MANTO: (These will all be short questions and short answers.)

BILL HARRIS: The threat assessment is from a military background the maximum hazard for which you need to design your requirements. That would have led to EMP protection; we would have had it long ago. Risk analysis involves reducing the expectation of harm and the need to protect as a function of the probability times the hazard, et cetera. And then you end up with incentives not to invest if you can't get short-term cost recovery. So that's where we've been, from my perspective.

__: Chuck, there's a question up here.

GEORGE SELLERS: Hi, I'm George Sellers. I'm with FDA [1:53] politics and social science [1:54] My question is, can you compare and contrast some of the dilemmas that the government is facing with this current administration coming in to having to maybe solve this problem without being such a mediator or trying to moderate a problem? What's the problem with people who might have concerns with government? Is it [2:14]? That might be a problem. Can you explain what the government can do to help people with their own problems?

__: I think in traditional thinking, gridlock is sometimes not a good thing. It's slowing all the process, it's slowing all [2:28] from getting passed in the law. But the issue of [2:33] time sensitive. That can be– that's the main issue right now, is that we have a process that isn't allowing us to move through quickly on this issue. And with the new administration, he's already laid out his update to focus on infrastructure and security within that. But also you have to realize that since is 1997, we've had no federal legislation passed through the US Congress. And that has been pretty much a stovepipe in the Senate. So if you're going to focus on an area within the federal government, I would say focus on the Senate right now in terms of legislation.

But then I think agencies – DHS, FEMA – organizations like InfraGard definitely play a role in getting information from that government domain out to the public sector, out to the individuals on how they can be resilient, how they can be forward thinking in protecting their own homeland of cities and counties, technical infrastructure at a local level. Disasters are traditionally thought of as local first and then it escalates up, to the county, to the state, and then to the federal.

And so, it needs to be a focus. And again, public opinion often drives public policy. So education, education, education, making sure you're having the right message to reach the government domain [4:01]

[Panel Member]__: Quick comment to that. So we're talking about the private sector. This is what we discussed yesterday. With the private sector, it comes down to what?

ALL: Money.

__: The bottom line, right? So we have to work on our message to tell business, and this is why it's a great opportunity with this new administration coming in, they're businesspeople. Why is this in your best interest to become more resilient? Why must we do these things to help protect the overall infrastructure of the nation? So I think we have a great opportunity. We have to work on the message, get that out there and sell it.

__: I'd like to add to that. There's not enough government to go around to solve this problem. There's just no way the government's going to solve this problem. We've spent a lot of time looking at it from a government perspective. We must look at it from a private sector perspective, to include it from a community basis. There is an emerging market and interest in sustainability. It's not that much hard to then bolt on a little bit of resilience into the growing sustainability movement across the country.

RAJ WILLIAMS: Hi, my name is Raj Williams. Full name is Rajeshwari Williams. It would take another half-hour to explain that, but anyways. My company is Green Sensitive Design. And I'm part of the Maryland InfraGard, and it's an amazing group of people, very smart. The big issue that I'm seeing, having worked in government for 18 years with a planning policy background from Clemson University, having worked in government, and having been in the hardcore city planning, that sort of things and seeing the disconnect.

For example, you have planning that goes with zoning, and you have it codified, you have rules and regulations, and all the developers have to follow that, and people in just general communities to follow that. Now we bring in cyber issues, and people are just trying to get a handle on that issue, and now we have EMP. People are like, it's oversaturated.

__: All right, Raj, where's the question in this, come on!

RAJ WILLIAMS: No, no, no, I'm getting there. The point is how do you get this EMP issue, to get to the hardcore community planning, whether it's zoning, whether it's policy planning. That is where the rubber meets the road. This is maybe [6:22]. You are preaching to the choir, and we all understand it's a big issue. But it really needs to get to that point.

__: Getting back to the communication strategy, we need to thread this in at all levels. Just not government, just not a few people who are interested in this type of topic. This has to be embedded and inundated throughout our society. So we need to talk that. We need to develop that within our leadership and push forward to make sure that we have the buy-in that we need.

__: Also, it's a portfolio management dilemma. It's not just EMP, it's not just cyber. It's not just blended neighborhoods, it's not just just-in-time manufacturing. It's a portfolio that can be managed. And the way to ease in to this in the talks that I've been involved with, there are blighted areas. There are blighted areas in Detroit that are very famous. So there are opportunities for some other communities to– but don't lead with EMP dilemma, you can't lead with that. People just shut down.

So, here's some areas that need some restoration. And by the way, while you're at it, while you're putting a microgrid up, while you're installing solar power, while you're looking at local food production, let's also consider, while you're doing all of this, let's look at the triple threat of EMP, solar weather and cyber.

That's a very different conversation. It's not a shocking conversation, it's an easy one to ease into. So what I recommend is, [7:47] have these conversations, back into this problem. Allow it to be an enhancement to a larger solution.

DAVID WINKS: This is David Winks with AcquSight. Just a question on the government side. Has there been any regulation or legislation proposed that would allow power companies to write off their investment for EMP protection directly [8:15] or to work with public utilities commissions to change the rate of return on investment so they get a much higher rate of return investing in EMP protection.

__: I know that's been a discussion that's been had at both state and level. I know there's been some legislation. I don't necessarily the specifics on what's going on at the federal in this. But there's optimism from the state levels [8:40] especially from the federal level coming down in the cost recovery [8:45] opportunity to do that differently than other states that don't have their own self-contained grid. But like we said, it comes down to money, when you have business-minded folks, and that's the first question they have. I have people tell me: I live eight hours away. Can you tell me how we're going to pay for this?

MARY LASKY: One of the things that happened in the book is that we have an Appendix Eight, and in Appendix Eight, Bill Harris has analyzed a lot of legislation that has gone on. So Bill, can you comment on that?

BILL HARRIS: Very quickly. First, through the legislation, their tendency is what's to be done at the state level. But first I'd say there's a bill that passed the House and the Senate, S.2012. It may go to conference in December or it may die or it will come up again. It has a provision that will allow states to add resiliency and charge through rate-payers for the distribution system, the transmission or distribution system. So that's the first time. It used to be dependent on FERC.

Secondly, at the federal level, under Federal Power Act, Section 205, FERC already has jurisdiction. It already allows black start contracts where one generator gets paid to stand by. And for the transmission system, those costs, and the fuel that black start in other facilities. So we already have that.

And the EMP Commission, under Section 1089 of the Defense Authorization Act of 2016 requires that commission to report back on whether they should get priority for defense facilities to states that have higher reliability.

So we already have some of that. Tax credits are a possibility; that's been done for solar. You could add an EMP requirement to keep some of those tax credits going.

I'll talk later on cost recovery.

CHUCK MANTO: Before the last question, I thought I'd mention something that reflects on the question that you've asked on legislation. At 12:45 today, Congressman Trent Franks will address us, or 12:50. He just did a vote in Congress, so he's going to give us an update of what's happened as of a few minutes, to give us an update there.

So through lunch, in addition to networking, be aware that I may come up to the podium and introduce him while we're finishing lunch, to be able to squeeze in some time to get updated.

Also later today, when we talk about next steps at the end of the day, we'll have a chance for more questions. So those of you who don't have questions answered now will have a chance to do that later in the afternoon.

With that, I will take a final question and give someone a quick remark.

By the way, anyone who is involved in the afternoon with slide presentations, make certain that you see Janet and make certain that your slides are given to her. And we have a question, the last question.

JOHN LOCKHOT[?]: Because I'm the last question before food, so I'll keep it short. John Lockhot from the US Department of Agriculture. As you move forward with your updated documents, I'm wondering, do you anticipate any linkages between the InfraGard special interest group and the various Sector Coordinating Councils under the National Infrastructure Protection Plan? For example, you mentioned some concerns about food and agriculture. I'm wondering if there's an opportunity here, maybe it's been done already, to network with private sector components of that food and ag Sector Coordinating Council and others on this issue? Thank you.

__: I think the quick answer is yes. It is very important. We talked about that a little bit yesterday in our discussion, too. We need to do a better job of pulling in the Sector Coordinating Councils. So we'll plan on making an effort to do that.

CHUCK MANTO: You're hereby appointed as a liaison to the Department of Agriculture [simultaneous conversation] [applause] And they will talk with you about that later so you can join their weekly calls on Thursday, usually at 2:00 in the afternoon. But if there's a reason why your boss or your boss's won't let you do it, they might even adjust it for you.

[Panel Member]__: It's a good example of how we've all been [13:10]

MARY LASKY: One of the things that John has said that these four people here are leading the press for Volume Two, and so if you could possibly join us, that would be just great. So please contact them so that you can get involved. Thanks.

CHUCK MANTO: And one last comment I'd like to say before we break in all of this, because I was supposed to talk about implications for best practices. [13:37] do that in about 30 seconds. So once you realize that this is a very complex area and if these people represented every body in the world that did every regulation whether it was a fire prevention code, a safety code, an IAAA spec, and all of the others, and throw the Europeans in the mix, they would probably not have enough brain power and authority to fully coordinate everything that we need to do if we would try to regulate this.

So one of the things we want to do is emerge the question of the community at large to understand how would we do a best practice from an umbrella perspective to sort of coordinate all of this in ways that haven't been done before.

And we can create emerging best practices in that arena, and we may even be able to develop training for that. We might be able to create a certification program for cross-industry best practices in light of these high impact threats. Nobody's really taken responsibility to do that. And this organization, in linkage with all of the other coordinating councils and so on and the private sector could marshal those who represent those points of view to be able to help us think through how we actually learn to be able to coordinate this so we have the best practices doing just that.

Because today, the best practices that you will see, including this afternoon yet, you will see some marvelous new technology, some best practices as it exists. We're just beginning. And we really need all of your involvement to get to where we'd really like to go.

I see a question from the panel. How could I not let them ask a question? Go ahead.

[Panel member]__: I want to put something in perspective right on your best practices. My boss [15:38] saying sometimes the best act of government is to get out of the way and let the private sector lead in some of these issues, especially when [15:47]

CHUCK MANTO: That's why I really want you to facilitate all the private sector, with public sector folks in the room, and that includes people who are inventing next-generation technology, with people who are managing the current technology, and anticipate where things are not only today, but the next two, three, five, 10, 40 years from now.

So again, thank you for all of this. There'll be opportunities for Q&A as we go along. But if you eat fast, talk twice as fast, out of both sides of your mouth, you'll get four times the networking done in a very short period of time. And thank you all for being here. [applause]

Policy Studies Organization and EMP SIG Relationship

Dr. Paul Rich

CHUCK MANTO: What we're going to do this afternoon, let me give you a head's up, Congressman Trent Franks is on his way, a little detained, on his way here, he's coming. What we're going to do is start first with some words from Dr. Paul Rich, and then as soon as he's finished, I will introduce the next speaker.

Is there an Andrea Boland in the house? Andrea Boland, wave, hello. Okay. Andrea's going to help the back of the room with my brother Sam to be quieter so we can begin to start. And you just drafted her.

We are here in large part because of the generosity of an association of universities that do public policy; many of you may not know them, the Policy Studies Organization. They do dozens of these conferences around the world, and do lots of publications. And we're very honored to be affiliated not only with the FBI, but with a set of universities that look at these problems. And today, you are in the building that houses their annual technology policy summit, called the Dupont Summit. And we've been the beneficiaries of their work and their generosity to be here today. And I would like to have this moment given to Dr. Paul Rich, who runs that society, and is the creative genius and force behind it. Dr. Paul Rich.

PAUL RICH: Just briefly to welcome you here again this year. This is the ninth year for the summit. Chuck is quite right that my membership represents universities all over the world. A conference coming up in Paris, co-sponsored with the French government in the Bibliotheque Nationale, and a conference here in March with the University of Wisconsin on the Middle East situation. And we just finished, we were over a few weeks ago at Oxford for a conference with Oxford University on the problems of Internet policy.

So we intersect with what you are doing, and have great enthusiasm for what you are doing, which is why we publish your papers and like to host you.

We ourselves go back to 1827. Our oldest journal is one you probably know, *World Affairs*. And we haven't seem to have solved problems, even though we've been working on it since 1827. I have to wonder about that.

But this house is extremely historic. And if you pick up or have picked up the program of the entire events today, there's a little history of it in the program. Just to give you one anecdote, Eleanor Roosevelt used to come here regularly, I guess to escape Franklin, and have tea upstairs in what we call the library. And it was here, while she was having her customary sort of reception for her friends and having tea, that she unfortunately learned of his death at Warm Springs.

It's a building that welcomes you and society, which I have the honor to be president of, that welcomes you every year, and we are honored to have you here. We're terribly grateful to Chuck. He puts a tremendous amount of time into talking to us and working on things, an absolute workaholic. He has built up a lot of friends in the university and the academic world.

So again, if there's anything we can do to oblige you, not only today but in the future. You are welcome at our conferences. And on behalf of the learning societies and universities, I am glad to welcome you here today. Thank you. [applause]

Best Practice Needs for Disaster Medicine and Public Health

Dr. James J. James, Executive Director, SDMMPH

CHUCK MANTO: So what I'd like to do now is introduce briefly our next speaker, Dr. James J. James. He's going to tell you a little bit more about himself, and we may be interrupted at any point in time, and here's how the interruption will work: If the Congressman comes while he's here, the Congressman will wait till he's done before his panelists come. So that may happen at any moment. And unless he comes before we actually have you start, Dr. James is starting momentarily. Come on up here.

Dr. James J. James is interesting. Many of our folks who are involved with EMP SIG's advisory panel for the medical industry know him well. If you've been involved in the American Medical Association, he has a long history of involvement. He will give you a little bit more about that, but I thought I'd make it applicable to what we're doing.

As you know, InfraGard is just 20 years old, EMP SIG just five years old. And a lot of the things that we're doing, all the groundwork, the nature has not really been done by others you might think could have done it. And the American Medical Association has not always put as much emphasis, from what I understand, to the disaster medicine community, and even to a lot of the public health community.

And so, many of the top people in the country who represent disaster medicine and public health have formed a group of their own called the Society for Disaster Medicine and Public Health that he founded and he chaired. Just from brute force imagination and no money, I assume, no budget – anybody familiar with that problem? So how do you start something from nothing, and the problems that you're addressing are some of the largest in society? So you'd think that the largest problems that face us would be managed by the people with the most money, but it's not usually the way it works.

So we have a very much kindred spirit here. And Dr. Terry Donat will be on the next panel. We've worked together this past July, actually, to put a tabletop exercise workshop together based on the triple threat work many of you contributed to last year.

So we think in the same way that we're working with the Protective Security Advisory program at DHS, we would like to also integrate more thoroughly with his society. And in the [2:26] proceedings you have, in the back you'll see an overview of the society, and you'll see a paper that he had jointly published that will give you a background on him and his society. So I expect that you're all going to read it. There will be a quiz in the morning for all of you who are seeing me next door Saturday for breakfast.

And with no further elaboration, I'd like to welcome Dr. James J. James. [applause]

JAMES J. JAMES: Thank you, Chuck. He made a couple of mistakes. He got the name almost right, James J. James, but he had it backwards. [laughter] Couldn't resist. I know. Anyway.

Anyway, I know the Congressman is coming. I do an awful lot of talks, but I do it to such disparate groups, I never quite know what to prepare. I've got some slides I'm going to run through quickly, especially the first half because I want to get to the second half. I think the one message that came through from just about everybody this morning is, you've got all these great speakers speaking to the choir. And myself and Terry Donat, we're talking about that. And how do you get the message to the people who need to hear it? More importantly, how do you get the message to them and have them act upon it?

And our society, disaster medicine and public health, we look at EMP, we look at electric grids, we look at outage, but we look at the whole spectrum of the disasters. And it's a terribly, terribly board spectrum. And the problems, regardless of what event you're talking about, seem to run through it. And that is getting the resources to do the job you know needs to be done, and finding the leaders in the private sector who will pick up the thing. It's just extremely challenging, and I'll leave it at that.

So let me run through the first set of slides. Or, set; it's maybe eight or ten slides. They tend to be basic in terms of disaster medicine, what we're about, what we try to accomplish, how we try to accomplish it. But the second part of it is the piece that I think you need to pay attention to in terms of how do we communicate with leadership.

Now, I'm healthcare, so I'm focused on the hospital and medical community. You're focused on all of those other sectors, including health. But my remarks are mainly within that [5:26]

Most people look at this and they go, Oh, boy, another story about 9/11. This poster actually appeared in the *New York Times*; I think it was 1962 or 1963. And why it appeared is a group of businessmen decided that the Trade Towers, which were on the drawing board at that time and had been approved, would be a disaster potentially, especially for commercial aviation. And this poster became the poster child for that group, and also a slide for the cancellation of the Towers.

But the important thing is, the individuals building the Towers paid attention to it and they tried to make structural changes that would allow the Towers to withstand this impact. And they probably would have, but what changed? Technology. The size of the plane. The amount of gasoline in the plane that hit.

So the thing is, you can't prepare for everything. I think too often we try to do 100% preventive preparation, when if maybe we picked 60% or 70% or a doable level, we'd at least get started.

This is just timeline of disasters. It's a little bit outdated right now, but the line continues up, up and away. And we're going to see that continue.

That's just another depiction of it.

This is the reason we're going to see it continue. Even if you had the same number of naturally occurring disasters, the impact on human population is going to go up exponentially because that's what's happening with the population. And as the population increases, food becomes scarce. Where do people live? They live in coastal areas, they live on the beach, they live near the water. And that's a continuing trend. And until population tapers off, I don't think you're going to see any tapering off of disasters.

The other important thing is, population impacts on every single node there. And those are the nodes that we're concerned with in terms of changes in the environment, the physical, biological, et cetera.

This is just a representation of cities at risk of natural disasters, nothing to do with nuclear, nothing to do with EMP. And the numbers to the right are not probability kinds of things; those are the populations in millions that are going to be potentially affected by a mega disaster. Tokyo, Yokohama, the planners there are convinced within the next ten years they're going to have a mega earthquake that's going to affect those areas.

This is what we're all about, and I don't want to spend a lot of time on it. I know it's basic. The folks in this room are mainly concerned with one part of that, and that's the mitigation and preparedness phase. I'm a firm believer, every one of us needs to be concerned about all aspects of the disaster cycle.

I'm going to have a little help here as I go along from Shakespeare. And we really need to have the wheel come full circle. Preparedness, we need information updated, we need to make a plan. You've got to put resources in place, but they have to be dual use. And I think it's you all, it started as we have a new administration, supposedly a big increase in infrastructure planning. We really need to build ourselves into it. Not just for electric grids and EMP, but for all of the natural disasters that we can too easily predict.

Better three hours too soon than a minute too late.

Response, early warning. Sounds good, difficult to do. A tsunami hits Japan, it travels 500 miles an hour and usually occurs within a 250-mile radius. So early detection is good, but not always possible.

One of the things we don't do when something happens is a thorough assessment. So we don't end up doing the right thing.

An overflow of good converts to bad. In Banda Aceh, right after the tsunami, the beach was littered with wheelchairs. The beach. Some organization had all these wheelchairs they sent over to Banda Aceh which lost 80% of its population. You couldn't even roll them in the sand.

Recovery, short term, long term. New Orleans is still recovering. We'll talk a little bit more about that here in a second. Not enough to help people out, but to support them.

The response phase gets all of the attention – the EMT, the search and rescue, et cetera. They run in, they do a job for a week or two, do it extremely well, and then they leave. The problems, the real problems are just beginning.

Mitigation. And again, that's where I think the biggest impact of this group and your efforts and policies go towards. And of course, the biggest problem is we build on the past; the past is the prologue to everything we do. And one of the examples here, and underscoring you can't prepare for everything, the great tsunami that hit Japan several years ago and ended up with power outages in the nuclear plants, et cetera. And people, "why weren't they better prepared?" They had a seawall in place that should have handled about 99% of tsunamis, by historical experience. That particular tsunami was a little bigger than what their maximum plan was. So again, without that seawall in place, think of how even greater the devastation might have been.

This is our vision. I'm sure it reflects the philosophy of many of the people here. And the word I was so happy to hear so often this morning from the speakers is resilience. Now, you were talking about community resilience and centers of resilience. I also feel very strongly we need to look at individual resilience. If you don't have individuals that are resilient, it's hard to envision a truly resilient community.

In the end, we want to end up with people who are ready, willing and able. Too much of our attention has been placed on first responders. You realize it's your fellow citizen, your neighbor, your family member who's responsible for more than 50% of retrievals in a major event for saving lives and limb. Does that mean the EMT is unimportant? Absolutely not. But it does mean we need to refocus more attention on our neighbors, on the public.

This is our publication. We don't write books; we have a professional journal, *Disaster Medicine and Public Health*. I'm going to leave a few copies up here if anybody wants to peruse it. We cover the whole spectrum, and we've had articles on microgrids and new technologies and power requirements in the field, et cetera. So it's not as if our community is ignoring what you're doing in the past, but it's only been since I've been working with Chuck and Terry that the magnitude of what we're looking at as a healthcare disaster has really, really become apparent.

Those photos were taken off the coast of Japan as the tsunami came in. And that was four photos taken by a single photographer who happened to be in position from the Medical Society of Japan. So we used that for our cover.

Now, here's the part– I'm going to turn a little because I want to be able to read. In terms of communicating to leaders, one is, what happened in the past couple of years that got more attention from our leaders than anything else in the healthcare field? Ebola. A real life-threatening event for most Americans. We spent literally billions of dollars on a non-problem. And I don't mean that, off the cuff, that it didn't have to be addressed, we didn't have to take measures to protect against it. We didn't have to support what was going on in West Africa, where it really was a devastating problem.

But there was no way you were going to have an outbreak, an epidemic of Ebola in the United States. And yet we were basically throwing people in prison; we called it quarantine. They used to do quarantine back in the Plague years. They had their militia from the Brits around the Plague houses and they were armed and you didn't leave your house. So our quarantine's not quite that bad, but all I'm saying is we spent an awful lot of resource on a threat that was much more political than it was practical.

Anyway, to get to leaders, you've got to get emotion. And what happened with Ebola is, everybody got emotional about it, everyone believed it, everyone was afraid. And I believe if you all are going to be successful communicating to leadership, you've got to get the message across on how serious a problem power failure is. I don't care if it's EMP. I don't care if it's a space storm or a nuclear, or whatever. What I think you need to focus on when you're dealing with the healthcare industry is power failure and the effects and potential effects of power failure.

People come back at me and go, Oh, well, we plan for that all the time. Yeah, you plan for three days. That's the requirement from Jacob[?]. And that means three days of fuel for a diesel generator. And you don't even have to have the fuel on hand; you can have a contract with the local gas stations to deliver that instantly. You all know in a true EMP event, all of that goes out the window.

The other thing, historically, they'll throw back at you: Well, we'll evacuate. Well, in your scenario there is no evacuation because there's no place to evacuate to even if you could accomplish it. We did our exercise in July on this. And after Terry finished and we were closing it out, on of the pundits in the room said, This is ridiculous because it will never happen. It can't happen. The probability is way too low. And we were talking about electric outage, not an EMP.

Well, guess what? There's a living example in the United States called Katrina as to what the impact of long-term electric outage is on a healthcare system. And I'm just going to run through it briefly. But I think this is a message that healthcare leadership needs to hear.

August 28th, mandatory evacuation. Unfortunately there was no mandatory evacuation of hospitals and nursing homes. Superdome became the refuge of last resort. Patients were still being admitted to the hospitals while the rest of New Orleans was being evacuated. That 2000 people, that was 2000 people inside Memorial Hospital, which we're going to talk about, who were there because they knew a hospital had power supply and food supply, et cetera.

So in addition to a couple hundred patients, you had 2000 people, family members, et cetera, et cetera, including pets, who were using that hospital for what you were talking about this morning, which is a resilient center.

August 29th, the next day, landfall, 135 miles. I don't know if you've ever been in a wind of 80/90 miles. You can't stand up, let alone when you're starting to get up around 135. Hospital was on generator power. Wow, that's good. Except generator power varies from hospital to hospital. The main requirement is that it keep the ER going, that it keep the ICUs going, it keep the respirators going, it keep the critical care areas going. Does it keep the air conditioning going? Does it keep the heating systems going? Not necessarily. These people ended up working in 120 degrees within the hospital.

Also, the hurricane warning came on a Sunday; this is a Monday. What do you think hospital staffing is early on a Monday morning? It was bare bones. Nobody came in. In fact, half the staff went home from this one hospital. So you've got one small group of medical providers who decided their duty, their responsibility was to those patients in that hospital at that time.

By Tuesday, August 30th, everything was under water, and they began to triage. Triage under the most extreme conditions. Generators failed. What does that mean? You're working in the dark. You're taking care of critical ill patients whose electrical support systems are no longer in place. You're doing it in the dark and you're doing it on 36 hours' no sleep.

So reverse triage. What's reverse triage? It means you do what you can to get the sickest people out initially. That's the first thing you do. This later goes back into regular triage where you have to do the most [21:33]

Quickly here, Chuck.

I'm sorry, that one didn't show up very well. This would show you the inside of what we're talking about, and believe me, it's like a third world country.

__: It's perfect, it's in the dark. [laughter]

JAMES J. JAMES: Right, we couldn't get the picture, it was in the dark. Never mind, I hate slides like this.

Anyway, inside, no electricity, no running water. The only communication was verbal and a few walkie-talkies. Temperature over 105 degrees. Limited supply of food and water. And no security. No plan for hospital evacuation. It wouldn't matter anyway; like I said, there was no place to evacuate them to. This was what was going on outside the hospital. And it was actually a lawless [22:38]

August 31st, a couple days later, the philosophy was basically to save yourself throughout New Orleans. They had a massive movement of patients to common loading sites. They carried them, two or three people per patient down to these loading sites, putting them on boats to go to areas where hopefully they could be helicoptered out or they could get them to the airport. One of the hospitals was using this drive-up thing where they used the garage and were able to operate cars and get people up to the eighth floor where helicopters were able to come in.

The hospital conditions, if you know what a special needs shelter is, that is something that is basically nursing support. It's not medical support. In this case, it was hospice. This was hospice support at this point in this hospital with 100-some patients.

And this is the scene I was mumbling about before which I thought was that other slide. And you can see, this is not the kind of medical care you expect in your local community hospital.

FEMA arrived three days later. They got rid of the airboats. There's good and bad to that. People, however, still would not evacuate without pets. You talk about other communities that you have to get into, don't forget the veterinary community. People do not move without pets.

Patients died. And the sad part is, many of those patients should not have died. They weren't intubated. Many of them were in that hospital for elective procedures or conditions that should have a fairly high probability of responding positively to certainly should have left the hospital. Many died. They were investigated. I'm not going to go into that aspect of the whole thing.

Problems. Hospitals and nursing homes are not required to evacuate, and they were used as shelters. That has changed since then. We have requirements that have evacuation plans.

The hospital was ill prepared. Of course the resources were overwhelmed. Remember, there were 2000 additional people using those resources. Lack of generator power. Generators were great for two or three days, or two days in this case, but there was no way to refuel them.

Mass casualty and triage, this is great in a wartime environment. What we saw in New Orleans, however, was not a war.

There is a duty to care. How long does a physician and a nurse expected to stay with dying patients, especially after they've been up continuously without sleep, going on three days, a skeleton staff.

What do you do with patients who cannot be evacuated? That was probably the toughest ethical question. And that's one that every hospital executive needs to be able to face, given an EMP type of event.

Hope turns to anger, especially in the intensive care unit. If you ever tried– I know you didn't, you're not there, but trying to keep a patient going who needs intensive care without support systems is physically demanding, to say the least.

There was no good solution, there were only bad choices. And that's the way New Orleans was in reality when they had a loss of power for a long period of time.

This was not New Orleans, this was New York Medical Center during Sandy. You can't see it very well, but what they're doing is evacuating a patient by hand down the stairs. And every once in a while you get a 400-pound patient and two people on a litter going down stairs in the dark. And that's reality. And that's what an EMP will result in.

Legal actions came later. We're all afraid to do the right thing anymore because we're going to be sued. And that's another aspect that's beyond what we're here today; just want to [27:26]

This is just a parting comment. Reese Winton [?] was actually supposed to be here today, but I had to cancel the meeting we had for yesterday and he didn't make it. So the future rests on us not allowing complacency to take hold. And that's what we need to overcome. And you're not going to overcome it with technical facts.

[applause]

Congressional Perspective and Critical Infrastructure Protection Act, or CIPA

Congressman Trent Franks

CONGRESSMAN FRANKS: [applause] I appreciate that. I so appreciate all of you. I know you've sat through quite a lot of speakers, and so I'm going to try to be fairly brief. But I have some good news for you today, but I'm going to try to tell you some of the bad news first, okay?

I'm continually grateful. I know there's a lot of people here that I don't know. I know a lot of people that I do. So I can't thank everybody. But I do want to thank Chuck Manto and all of his colleagues for organizing this event because you guys, I promise you, have coalesced not only for good reason, but you've have some profound effect. And I'll try, as I said, try to justify that statement momentarily.

I also want to thank Bill Harris. Where is Bill? There he is, sitting back here. We worked on the *Powering Through* project openly. I think it will be families and local and state governments who hear and bear the brunt of this responsibility. It certainly was outlined in the last speech fairly well. Indeed, that the folks in the local spots that will have to deal with the picking up the pieces if there is a tragedy.

Of course, the book I think is tremendously helpful in sort of preparing the people for the unthinkable. But it's a powerful piece of work.

I want to thank, I saw him back there, Congressman Roscoe Bartlett. [applause] Congressman Bartlett really carried the torch on EMP for so many years. And I truly believe without his foresight and dogged determination to this fight and doing the right thing in sometimes the face of almost relentless and obstinate refusal by so many to even recognize the threat. I just believe that without Congressman Bartlett, the profound advances that we've recently made would not have been possible.

And I just perhaps finally would recognize all the members of the EMP Commission for their invaluable work and what they're doing. We were able to get that thing started again, as you all know. And Dr. William Graham and Dr. Peter Pry, especially, have been working tirelessly and thanklessly sometimes with Congress and this DoD bureaucracy to safeguard the United States from this existential threat of a potential EMP attack. And I want to thank both of them especially for everything that they do and that their colleagues have done to protect America's vital national security.

And for many reasons I believe most of us in this room are more optimistic about the future than we've been in a long time. And despite some of the laborious natures of this task, it appears that Congress and the federal government are finally coming around to recognizing and doing something about the EMP threat and our vulnerable grid.

That said, you all know that the threat is a growing and remaining threat. You know it's real. And it is especially, I think, exacerbated by this disastrous nuclear deal that President Obama has made, along with the influx of cash to Iran. And this is to an enemy that's keenly aware of our weaknesses and the dangers that we face here. We know the Revolutionary Guard was able to successfully hack into the computer-guided controls of the Bowman County Dam. And that's again some time that they did it, but we were finally able to know about it.

And given the daily proclamations of the mullahs about the Great Satan, they continue ballistic missile tests and ongoing nuclear programs in Iran to ensure that, unless something changes, Iran will one day be capable of exporting the Islamic revolution to the skies above the United States by detonating a nuclear warhead over America to catastrophically damage our grid.

The North Koreans, more perhaps erratically, and perhaps even more horrifyingly, because they're more capable than the Iranians in many areas right now, they're more advanced in some areas, have conducted five nuclear tests in the past decade; four of them under the Obama administration, and two of them just this year. And I can just assure you that this threat composed by this hermit kingdom, they call it, is deadly serious. And the defense and intelligence communities are reassessing and upgrading the threat assessment even now.

The situation grows even more destabilizing and troubling in the light of the North Koreans' relentless pursuit of ballistic missile technology capable of hitting the United States. While our news media mocks the Kim dynasty, there's growing evidence that North Korea in the not too distant future may develop capability to detonate a nuclear warhead over the American homeland. They've already talked about this in terms of being able to potentially hit Seattle in the next five years. The outgoing director of the CIA mentioned that.

And they're orbiting a satellite over this country right now. Again, I know I'm saying things that most of you know, but to be able to launch and orbit a satellite over the United States really demonstrates the critical capability to do the worst to us because the payload of a satellite, if it's translated into a neutron-based warhead, it doesn't have to be a big one. The way all of the things match – the trajectory, the missile technology, the launch – all the technology needed to attack us they have. And that's a growing situation.

And I know that some believe that the normal deterrence and mutual respect for our nuclear arsenals might be lessening the threat from Russia and China, but those two countries are still nonetheless working tirelessly to exploit the gaps in our missile defense network, which could leave us vulnerable to an EMP attack.

And finally, the stateless terror groups and the so-called lone wolves inspired by their own ideology are increasingly aware of the potential weaknesses of our grid and the low risk/high reward cost/benefit analysis of attacking. The Metcalf station attack, all of you are so much aware of it, where with rifles and wire cutters they cost us $15 million to try to fix that.

And the cyber caliphate is alive and well and ever virulent. And Russia has demonstrated tremendous capability to infiltrate the electric power grid, the interconnectedness of our own technology and the increasing opportunity for the world's power grid to be hacked.

In spite of all of this, and this is really what I came to talk to you today, a lot of this is prepared information. I wish members of Congress weren't so busy to where we could actually sit down and slow down and develop the kinds of message that we want to give a group like this. Because I can't express to you the level of respect that I have for each one of you. Frank Gaffney back there, he's been in the middle of this so long, and I see Chris. I mean, if I start naming names, it won't make sense.

But I will just tell you, don't think that because we're running all the time that we don't respect what you're doing. And I want to give you some what might be encouraging notes. I know some of you are aware of it. You all know that the Congressional EMP Commission has been reformed and it's begun working. And they're facing a great deal of pushback from the DoD bureaucracy. And we're not done fighting with the bureaucracy. We're winning that fight and we're going to get them taken care of.

But the commissioners have begun working, and that should be extremely encouraging to all of us. And bureaucratic hurdles from the Pentagon notwithstanding, we anticipate, I do, extremely enlightening and timely report as early as next summer. And I think that we can expect this report to serve as a guide for both Congress and the new Trump administration in mitigating the threat to our grid.

And it is a special privilege for me to be able to announce to you today that the language from the Critical Infrastructure Protection Act, or CIPA– how many of you know about CIPA? Well, CIPA made it into this year's defense bill which passed the House this very day. [applause]

Section 1913. Among other things, CIPA, you all know, it creates a statutory requirement that six months after the enactment of the NDAA, an intelligence-based review will be conducted to review the risks and consequences of EMP and GMD facing our critical infrastructure and submit the findings to Congress. But more importantly, and most significantly, these findings must include a recommended strategy to protect and prepare critical infrastructure of the homeland against threats of EMP and GMD, and update it every two years.

CIPA, under the Trump administration, will ensure that the Department of Homeland Security and the appropriate agencies and utilities are accurately assessing EMP threat and taking the appropriate actions to safeguard against it.

I promise you, this is a tremendous and unprecedented victory for all of us. This bill will pass the Senate, and I believe the President will sign it. We have had really our first huge beachhead of major legislation on this issue. And I say this with the most desperate sincerity: it's because of people like you that stayed with it when other people sort would have rather have ignored the issue and continued to stay focused and alert and alive.

You know that famous philosopher Bruce Lee said [laughter], The difference between a warrior and an ordinary man is a laser-like focus. And you folks have had a laser-like focus and we have now something very significant and that I believe the effort by the new Trump administration to have a major infrastructure buildup should dovetail. There should be tremendous synergy here, to begin to look at ways to protect this grid as we go.

And all of you know that when you're working on an engine, you might as well fix things that need to be fixed while you've got the thing taken apart. And it's important to realize that we really have a major victory here.

And finally, the Shield Act made it into the House energy bill this year, but unfortunately the energy bill probably will not pass Congress this year. But that means the wind is now at our backs for the next Congressional session when we take this back up. And I believe that there is a good possibility that the Shield Act will either be passed as a standalone in Congress next year or it will be passed as part of the energy bill. And that's a big deal, because folks I've got to tell you, if we get both CIPA and the Shield Act signed into law, the job of at least Congress related to EMP will be essentially accomplished. And it will be up to all people like you, the administration and others to implement that legislation. And we will be going down a path that is set on a mandated mindset to go forward and fix this thing.

I'll just say, by the grace of God, this sought-after moment when America will no longer be existentially vulnerable as a nation to a widespread, weaponized electronic magnetic pulse attack might finally be in sight. And I just want to thank you all for being at this summit and for staying committed to what you're doing. And with so many working so hard, not only are we in a better place than we've ever been, but I really believe that we have an opportunity to go forward as we've never had before.

And I thank all of you with all of my heart, and let's charge the gates and make sure that some day, as I've said before, the people will be watching TV or outside playing baseball or something and an event with such magnitude will have an effect so small that only a few people like you will know how much tragedy has been averted.

So God bless you all, and thank you very, very much. [applause]

CHUCK MANTO: Congressman Franks, thanks. We have to make a quick presentation with Mary Lasky [13:20] came out of the *Powering Through* book, which you gave a wonderful endorsement to on the back cover; we appreciate that. And there's your copy. Here is another copy of something that includes your work, along with us, from the last year's conference proceedings. And this is a memento from the 20th anniversary of the InfraGard organization in honor of your leadership efforts in this regard. Thank you very much.

CONGRESSMAN FRANKS: Thank you all very much. A little light reading here. [applause]

Medical Panel – Best Practice Assessment Underway by DHS, NIHS Hospitals for Resilient Hospital Microgrids

Dr. Terry Donat, MD; Dr. Earl Motzer, Dr. James J. James

CHUCK MANTO: Now what I'd like to do is, we'll have a panel with Dr. Terry Donat, Dr. Motzer and Dr. James James. I think he's back, there he is. I know Dr. Motzer is heading out for a plane shortly. And so, what I'd like to do is, I'd like to introduce him briefly.

He's on the InfraGard board. He's been the CEO of a hospital for roughly 50 years or so. He's been involved with HHS Coordinating Council. He'll explain all of that in the few moments that he has. And he may want to make some initial comments. And then to the extent that he can sit back down and join us for a couple minutes more, you'll understand that he will have to leave to catch a plane because we're a little bit behind.

But Dr. Motzer, why don't you begin. And you can either come here or sit there, whichever.

Dr. EARL MOTZER: Thank you very much, Chuck. Is this on, can you hear me? Today I'm sitting here as retired hospital chief executive officer who was responsible to a board of directors and the community for the safety and security and resilience of my clinical infrastructure, my technology and my horizontal and vertical people inside the infrastructure.

I would say to that hospital CEOs like myself know that there is something like EMP out there, but we don't understand it. And the seminars we've gone to in the past have been technical in term. And I was lucky enough as chair of the health and public health sector, in answer to the gentleman from agriculture, yes, some of the 16 sectors are definitely involved, through a grant with the support of Chuck and with the support of Dr. Jim Terbush to do two things: One, to educate healthcare and public health leaders in non-technical terms – in other words, EMP made simple – so that they have a better understanding of what it is. You all know that you have fear of the unknown, but if you were able to educate my peers about this, they will know better how to act. And so, we will do a handbook, we will do a website as phase one.

Phase two, through Chuck's involvement with DITRA, we will be able to fund the assessment of three hospitals. It doesn't sound like much, but it's a start. So that we can see and determine whether or not they can achieve at least grid parity; meaning that they would be able to get microgrids, EMP protected, and they would be able to sell electricity back to the utility companies. And the proceeds from that would offset the cost of the microgrids.

Now of course, Chuck, as imaginative as he is, hopes to be able to find money that we can do that without even worrying about the parity. And hopefully we'll be able to make that happen. But through the courtesy of Deputy Assistant Secretary Bob Kolasky, who works for the Assistant Secretary who was here today, he was able to convince Congressman Hal Rogers to put aside $3 million for ten challenge grants. Bob previously worked in business and industry, and understands the value of challenge grants to people within organizations. And Congressman Rogers went along with that.

So there are ten grants, but there's only one related to EMP and healthcare. And thankfully, I always know enough to surround myself with people who know a lot more than I do, and that's why I surround myself with Chuck and with Dr. Jim Turbosh.

So we have one year to be able to provide the workbook, the handbook, the website, and to be able to demonstrate to at least three facilities nationwide that it is possible to make this happen. And I know my colleagues well enough that if they see that that's happening and they can do it, why they're going to be able to hear, not too far along, from their board of directors if they're doing it, why can't we be doing it.

I liked Dr. James's comment about not necessarily hanging our hat on EMP, but the issue of the power outage because we're used to, as he indicated, for 72, sometimes 96 hours, to be able to survive. But after that, we're in trouble. And if we hang our hat on how are we going to be able to keep our doors open when we have a power outage, it sounds good to me.

So I apologize for having to run away, but I wanted you to know the good news, that one of the 16 sectors does in fact have a grant that has experts in a booth, and we intend to make this happen. Thank you. [applause]

Dr. TERRY DONAT: My name's Terry Donat, and I'm secretary for the InfraGard program in Chicago, as well as I co-chair the healthcare industry advisory group for the EMP SIG with Jim Terbush up in Colorado. One of the things that I got involved in a few years ago was as emergency manager and looking at not just EMP but electricity and infrastructure support for hospitals.

As a surgeon, we take it on a daily basis, I can't do what I do without the whole cadre of people [6:42] a dog's head shape wondering why medicine has to do with security, particularly national security. And if with walls and guns and guards [7:01] medicine and a bit more than that. It's, I think, defining security more as the intelligence [7:10] awareness and the coordinated actions that give a survivor advantage, whether it's the individual or a community. I think that's what you really have in medicine on a daily basis.

So when we look at who's attending these events– Jim Froist[?] is here, he's a dentist from Pennsylvania. We don't have any veterinarians in the crowd, but I'm sure we will going forward, using the agricultural importance in security. And myself in medicine, I think one of the big things that applies is that medicine is not just on the back side, if anything it's on the fore side, in that you can't preserve those capabilities as security capabilities if you don't address the infrastructure.

One of the things in looking at what this panel is about is assessment for microgrids for hospitals or capabilities for hospitals. And although James James and I have talked a bit, and Earl talked about this, capabilities, one of the things I want to mention was, in looking at an assessment, one of the things I think gets completely left out of most hospital exercises is loss of communication.

Hospitals' communications is very strongly either internally secure provided they have power, or externally it's always based upon infrastructure communication through either fiber optic lines, Internet lines, or other means. And the only really radio communication you have is with EMS. If you were to ask one thing of anyone in this room to do when they go back to [8:48] and assess resilience in their hospitals, ask your hospital, with every capability intact, all power, all water, all supply lines, pharmaceuticals and all their staff, how would you communicate with your staff externally, with other agencies externally, or other hospital supply companies in the absence of infrastructure communication. And I think you would start to see some lights there. Because those are actually very easy things to surmount amateur radio or other non hard-wired infrastructure-based communications, both at local level, state level and the national level [9:26]

But I think one of the things that I would impart to you, as Joe and I were talking, a lot of these conferences we tend to revisit the same issue without talking about what should I do when I go home. It would be one thing I would impart to you to do immediately.

I think one of the big problems in medicine that medicine has a hard time wrapping its head around is dealing with loss of electrical power. We are not only so reliant on it to do our daily work, but we believe that it will always be there. And we also believe that it will be there for our families when we go to work and when we go home because we're very infrastructure fixed based facility in general except for EMS. So that leads us to believe we tend to focus on our own locality and our own facilities. We tend not to look at how infrastructure impacts or loss of infrastructure impacts those.

A secondary thing is that in trying to comply with a thing like the new emergency preparedness directive from CMS whereupon our emergency preparedness efforts are contingent on [10:27] participating and getting paid through CMS, what happens is emergency managers and hospital administrators very much focus on the details of their fine processes rather than the large impacts that could affect them externally in the whole hospital.

Other initiatives that really brought our attention are things like antibiotic [10:46]. There's a lot of time and effort being dedicated to those things, and they're important, but they will mean very little if you have loss of infrastructure inside the hospital. [10:58] some basic capabilities.

And then, I think one of the things I wanted to mention, I've been very lucky to be able to support Chuck and Earl Motzer through the InfraGard program coordinating committee, and James James in running and having forums in which we look at these issues. One of the things I wanted to mention, Dr. Motzer had to leave, but the sector coordinating committee is the private sector side primarily for DHS coordinating activities, along with the government coordinating activity. And through the sector-specific [11:33] developed last year, healthcare is the only sector, apart from energy, that mentions space weather and other grid threat activities [11:45] that could affect it.

So currently, we're the only two sectors that have a plan for recommendations for planning in those guidance documents, which we've been very fortunate to have primarily through the efforts of knowing what Bill Murtagh and the Space Weather Prediction Center was working on with the White House prior to actually starting and initiating that. So that's been very successful.

Two other things I wanted to mention. Last year, James James asked me to help with a tabletop scenario we ran through the Society for Disaster Medicine and Public Health [12:17] And we did that this past July. And some of the things that came out, we ran a tabletop scenario with a space weather event impacting most of the Eastern connection and then the area around Seattle, much akin to the conditions that Metcalf [12:34] at 50 degrees.

Then we also put in, at the end of the first week, an opportunistic attack upon large transformer field that would have increased the area affected. And we were fortunate enough to have Margaret Dunn who was from, at that time, Georgetown University Internet War College from Pennsylvania down to talk about some studies they had looked at. And what would happen in just five minutes, in the first week and the first month of not having access to insulin or having access to electricity or refrigeration. And what you see that almost all major diabetics, certainly all the Type 1s, and most of the Type 2, meaning they're insulin-dependent versus non-insulin-dependent but take other medications, most of them don't survive more than 34 days. And that ends up being millions of people in the eastern half of the country.

So you're looking at dramatic impacts very quickly. Also we had Chris Finder[?] from Health and Human Services, because there are lots of GIS resources where individual counties or towns or states can look at grid-vulnerable people who are energy dependent where there's oxygen, people on dialysis, insulin pumps or other medical devices. And that's going to be increased with the Internet of Things, as well as being pushed [13:51] the last decade to get people out of these [13:53] facility into the [13:56] Those people, when they run out of power, they end up in hospitals and hospitals in general don't have enough electrical capacity to do anything in actual wattage [14:09]

The things that we really realized out of that work, communication will be a big problem because you can't coordinate if you don't know what your situation really was. The other thing is that maintaining a staff after the first week is going to be virtually impossible in most major urban centers. Not just because of lack of transportation, but people's ability to deal with what's happening in their own responsibilities with their families and with their neighbors and those requirements. And so, pretty much after the first week things go south primarily in healthcare.

The last thing is, while right now, for instance, we can do [14:47] plastic surgeon. I'm sure he does some cosmetic industry and kind of other things. The mainstay of what disease is is in this country has been not wiped out, but really [14:57] So we can deal with more of the aesthetic [15:01] what we have to do for medical care. One of the things we realized is that within a week, especially with things like diabetes, the natural history of disease gets revisited upon the population that has been affected. Meaning, when you don't have a way to medically intervene, hypertension, strokes, trauma, diabetes, all those things, the natural history of what happens to people comes to bear very quickly, within the time span, and there's no way to mitigate those things, or even support those people [15:28] care.

So what I really would say is that in looking at this, it's encouraged, if you get the chance to encourage any hospital administrators you come in contact with to look at this. Only because in a time of need, hospital administrators and medical staff are going to [15:45] for families, just like everybody else. So if you do not have plans in place, and I would especially urge alternative communications plans because you really can't start anything coordinated without doing that.

Thank you.

Dr. JAMES J. JAMES: I have a couple of things, but it would take about two hours. [laughter] I don't want to start coughing again, that's embarrassing. I really prefer question-and-answer, but I just want to point out one article in our journal which was "Tools and Checklists Used for the Evaluation of Hospital Disaster Preparedness: A Systematic Review." It was a very thorough review of all the literature looking at all the criteria for testing hospital readiness.

Electricity was mentioned – one word, one. And everything on sustainability had to do with three days. So there's a tremendous amount of work to do. As an organization, called JCAHO. They do accreditation of 80% of US hospitals. If you want to get hospitals to pay attention, you need to work through JCAHO, creating requirements. And why do they listen to JCAHO? Because without the accreditation they don't get Medicare reimbursement.

I'd rather open it up to questions.

CHUCK MANTO: We have time for one question if you have a microphone. But if there's no microphone, there'll be no question. [laughter] [17:27] highest bidder? 10 bucks? 20 bucks? Sold, 20 bucks! One question.

STU ROYTER[?]: A quick question. You mentioned the history–

CHUCK MANTO: Identify yourself.

STU ROYTER: Stu Royter, submarine officer. You mentioned the history of things that would repeat, like hypertension. How about cholera? How about typhus? Typhoid? Things like that that also are lurking in the woodwork

TERRY DONAT: Without a doubt. While we talk about the things that we [18:00] we're very lucky [18:00] to have of parasites and water-borne and things in this country. We have specific preparation mitigation [18:07] Those will all [18:10]

CHUCK MANTO: You're all welcome to stay for the next steps conversation because at least some of these people will be here, we'll be here all the way to five, five-thirty, as you can see in your program. So you'll have opportunity for more questions.

But I wanted to thank them because many folks in the country have never been inside the control room of a nuclear power plant or an ISP or a data center. But everyone's been in a hospital or walked by one, right? So if we can get this concept into the minds of the folks in hospitals, and have people see how hospitals can become sustainable, because they make 20-30% of their own power and store it locally so they can stay open indefinitely, instead of closing their doors when we need them most, they can then say, Well, if the hospitals can do it, why can't my institution do it?

So they're one of the top infrastructure plays anyway, but from the standpoint of an icon, they are the great American icon. And you guys are just that, so thank you very much. [applause]

Preparation for Best Practices for EMP-protected Microgrids

Amb. R. James Woolsey; DHS SHARES
Mr. Robert Lachenmayer, COO, Pos-EN
Mr. David Geary, Power Analytics
Mr. Douglas Kiesewetter, Carpe Solar
Mr. William Kaewert, SENS

[Editors note: the recordings after Amb. Woolsey's introductions have been lost and were not available for transcription at this time.]

CHUCK MANTO: –microgrids to come up to the front. That'll be Honorable Ambassador James Woolsey, Mr. Robert Lachenmayer from Pos-EN, David Geary from Power Analytics, Mr. Douglas Kiesewetter from Carpe Solar, and Mr. William Kaewert from SENS. [:19] chairs. I'll stand up here in case somebody needs the podium. And what I'm going to do as they arrive– we don't have a lot of time [:27]

I'm going to begin to do brief introductions of them so we can save another minute. First of all, Mr. Robert Lachenmayer is the COO of a company called Pos-EN. A great, high-tech start-up, but the people who are starting it up are not start-ups; they are industry seasoned executives from the utility industry, and some guys have actually gone up and said, Why have you gone over to the dark side and left us? They went from a few million dollars of microgrid contracts and handle well over $100 million or so, if I'm not speaking out of turn. So we're very glad to have him here with us.

Mr. David Geary is working with a group called Power Analytics. Lots of groups from the telecom industry understand how to manage phone networks and telecom now placing that technology over the power world, so you can understand how to understand, monitor and manager power grids. Very important work.

Mr. Douglas Kiesewetter from Colorado has a breakthrough new technology from his company called Carpe Solar – seize the sun – and he'll be, again, just brief comments about what he's going to do. Everyone's going to get four minutes or so, just play like Congressional testimony time. They give Congress that brief overview, and then maybe there's time for a question or two.

And then Mr. Bill Kaewert from SENS, Stored Energy Network Systems, or Stored Energy Systems, there we go, SENS; I've got to account for all those letters, there's four of them. He's also from Colorado and has a great power converter technology, among other things, to talk about.

And of course, what I'd like to do is have the challenge about how we can take responsibility to solve this problem right where we are, with our institutions, our technology, locally. We're going to get the challenge from Ambassador Woolsey to them about what it will take to make this work. Ambassador Woolsey, please start off and give us a grand challenge.

AMBASSADOR WOOLSEY: If you're one of those folks who makes fun of preppers and doesn't think that it's ridiculous to talk about aluminum hats, and so on, you will not like what I've got to say.

The two big kinds of problems that we face in life; one one might call malignant and the other malevolent. A malignant problem is one that is naturally occurring, whether it's cancer or some other. And we can work together and do work together to deal with these kinds of problems frequently through [3:23]

The other in a way is less common, call it a malevolent problem. A malevolent problem is if someone is trying to do you in. They're trying to take your property or kill you. And you have to fight back, and you don't get to define the problem as if it doesn't exist.

So in the world of EMP, there are a number of people who would like to only worry about solar weather – talking about folks in the government. There are a number of people who say well, it's like science fiction and Newt Gingrich wrote a novel about it, how can it be serious? Et cetera.

Let me just say this. In a number of circumstances in which you're trying to solve a problem, a technological problem, if you solve a single part of it, you probably want to start with the simpler and the smaller and grow over time so you can deal with one of the more complex.

But with respect to EMP – and I'm going to be talking here about detonations in lower earth orbit of nuclear weapons, not anything else – that's the big problem, that's the hard problem. But in terms of solutions to solar weather or this or something similar, in a lot of ways this is the easier problem to solve. And it is the easier problem to solve because the malevolent attack against us is something that, if we deal with it, we will have largely, perhaps almost entirely, dealt with the lesser included cases. We're not trying to start simple and small and build up; we're going right for the most damaging and devastating threat – the detonation of a nuclear weapon in low earth orbit above the United States, knocking out with two types of electromagnetic pulses our electric grid, including the transformers, and our individual electronic devices, both.

And if that happens, according to the Graham Commission, there are two possibilities for percentages of Americans who die in the first year. One, the moderate position is that only two-thirds of the population dies – from deprivation of food, water, electricity, communication and social unrest. The more serious estimate is that 90% of the American population dies in the first year, because of those causes. And particularly as social disruption gets worse and worse.

This is not the stuff of novels. There's some interesting novels written about it, but it's not imaginary. It's not something you can pass over. It's not something you can ignore. You have to face it, and face it early. Because if you deal with this problem successfully, both in building resilience into our grid and moving perhaps simultaneously towards distributed generation, in such a way that you can contain and maintain the ability for our grid to function, at least in part, while we move into being able to rely on something like distributed generation microgrids, and so on. If we approach it that way, we have a chance.

But when you read in the papers that this is not a serious problem because North Korea, for example, doesn't have a miniaturized warhead that would be necessary in order to execute an attack on the United States – nonsense. You do not need a miniaturized warhead for this; a small, simple warhead, indeed the warhead that was used over Hiroshima, would work. All you have to do is get it into orbit and detonate it, when you want, above the United States or any other enemy you have. Let's say you are North Korea. That detonation will do the job. Unless we have taken steps first to build resilience into our grid and make it possible to operate with developments such as microgrids.

There's no way to make it as hard to detonate something in orbit as it is to hit a target on the far side of the earth. It's not accidental that the first thing both the United States and Russia did in space was to orbit a small satellite; they with Sputnik, we with Explorer three months later.

That's because it's easy to do, not hard, comparatively speaking, targeting something on the far side of the earth; very easy. Is special work on miniaturization necessary? Nonsense. Anybody who tells you that does not know how space-based systems work, does not know how the aerospace systems have developed, and how they can be used.

So if you choose to stand aside over the course of the months to come, and hopefully it can be begin to be measured in months, not years, but if in the course of the months to come we can move toward distributed generation, we can toughen our existing grid a bit as we get there, then we have a chance within a relatively short period of time denying to North Korea, and I'm afraid probably within a year or two denying to Iran, the ability to destroy our civilization.

One does not need to have only a madman on the other side of this divide in order to worry. People will say, Well, the North Koreans aren't going to do that, the Iranians aren't going to do that, they're not crazy; we'd retaliate with our weapons and what could go wrong? Well, retaliation doesn't do it if none of your electronics work. And if you can't tell where the shot, the detonation of the satellite came from, where the satellite came from, because it was launched into orbit going around the earth on the south side first, so you can't tell when all of our sensors are focused on something coming at us from the north.

So if one has a detonation in space of a simple nuclear device without it targeting anything on the earth, without it having to reenter, without it having to have heat shields, none of that, just in orbit and somebody detonates it – North Korea or Iran – we would see the destruction of American civilization. And if you don't want to work that problem, if you don't like malevolent problems, you only want to work on malignant ones – solar weather – okay. But you are taking in hand a risk that if the North Koreans or the Iranians or any other country – before too long it may be terrorist groups with balloons that can get something high enough to detonate over a major section of the United States – if you don't want to work that problem, you are not working the real issue.

You are risking– if our civilization is destroyed, your inattention to the key issue has left blood of something approaching 300 million Americans on your hands.

Thank you. [applause]

Biographies of Presenters

Baker, Dr. George
Email: BakerGH@JMU.edu

Dr. Baker is emeritus professor of applied science at James Madison University (JMU). In addition to teaching graduate and undergraduate applied science courses from 2001-2012, he organized and directed the university's Institute forInfrastructure and Information Assurance (IIIA). From 1987-1994 Baker led the Defense Nuclear Agency's EMP R&D program protecting strategic systems against electromagnetic pulse (EMP) effects and developing DoD's EMP guidelines and standards. From 1994-1996 Baker directed DNA's Innovative Concepts Division, managing advanced technology development and demonstration including electric guns, high density energy storage, radio-frequency directed energy weapon technology, and space nuclear power.

A primary research interest stems from his experience as Director of DTRA's Springfield Research Facility, a national center for critical system vulnerability assessment and protection guidance, during 1996-1999. His organization was responsible for developing and implementing the Joint Chiefs of Staff Force Protection assessment methodology. From 1999-2001 he served as a senior scientist at Northrop-Grumman, advising DTRA nuclear effects and test programs. He now applies lessons-learned from DoD experience to critical national infrastructure assurance and community resilience. He consults in the areas of critical infrastructure protection, EMP and geomagnetic disturbance (GMD) protection, nuclear and directed energy weapon effects, and risk assessment for customers including DoD, DOE, DHS, the White House, National Guard units, the National Park Service, SAIC, and Defense Group Inc. He also serves on the Congressional EMP Commission, the Board of Directors of the Foundation for Resilient Societies, the Board of Advisors for the Congressional Task Force on National and Homeland Security and the JMU Research and Public Service Advisory Board. Degrees include M.S., Physics (University of Virginia) and Ph.D., Engineering Physics (U.S. Air Force Institute of Technology).

Bartlett, Congressman Roscoe G.
Email: roscoegbartlett@gmail.com

Elected to serve his tenth term in the United States House of Representatives, Roscoe G. Bartlett considers himself a citizen-legislator, not a politician. Prior to his election to Congress, he pursued successful careers as a professor, research scientist and inventor, small business owner, and farmer. He was first elected in 1992 to represent Maryland's Sixth District.

In the 112th Congress, Bartlett serves as Chairman of the Tactical Air and Land Forces Subcommittee of the House Armed Services Committee. Owing to his 10 years of experience as a small business owner, he also serves on the Small Business Committee. One of three scientists in the Congress, Dr. Bartlett is also a senior member of the Science, Space and Technology Committee.

Prior to his election to the Congress, Dr. Bartlett worked for more than 20 years as a scientist and engineer on research and development programs for the military and NASA. Nineteen of his 20 patents are held by the U.S. Government for his inventions of life support equipment used by military pilots, astronauts, search and rescue personnel, and firefighters.

In 2008, *Slate* magazine applauded him as "an advocate for reducing dependency on fossil fuels." The Association for the Study of Peak Oil (ASPO-USA) created the Roscoe G. Bartlett "Speak Truth to Power" Award in his honor in 2008. It had previously awarded him the M. King Hubbert Award in 2006 for his leadership in the Congress to promote efficiency and conservation and alternative renewable sources of domestic energy to enable the United States to overcome the challenges to national security and economic prosperity of global peak oil. Congressman Bartlett is the cofounder and cochairman of the Congressional Peak Oil Caucus. He is also the cochairman of the House Renewable Energy and Energy Efficiency Caucus and Defense Energy Security Caucus. He is also a member of the Oil and National Security Caucus.

Boland, Hon. Andrea

Email: sixwings@metrocast.net

State Representative Andrea Boland recently completed 8 years (or 4 terms) in the Maine legislature. She is considered a leader in safety issues of electromagnetic radiation, especially from cellphones and smart meters. She became involved in electric grid protection against electromagnetic pulse and geomagnetic solar storms (GMD) at the suggestion of her regular scientific advisor. Her work is supported by several national experts.

She has a B.A. degree from Elmira College and an MBA from Northeastern University, and studied at the Sorbonne and Institute of Political Studies in Paris. She was awarded the 2011 Health Freedom Hero Award by the National Health Federation for her work on health freedom and safety. Her legislative work has led to confronting major corporate interests on matters of transparency and regulatory capture, and public protections.

Breor, Scott

Email: scott.breor@hq.dhs.gov

Scott Breor serves as the Director for the Protective Security Coordination Division (PSCD) within the Office of Infrastructure Protection at the Department of Homeland Security (DHS). Mr. Breor is responsible for the day-to-day business and management of infrastructure security specialists known as Protective Security Advisors (PSAs) and the Office for Bombing Prevention. These programs provide communities with effective vulnerability and security gap analyses, training, support to Special Events, and improvised explosive device (IED) awareness and risk mitigation training. Mr. Breor has over thirty years of military and senior executive experience in the United States government. Prior to DHS, Mr. Breor was a Naval Aviator and had served as the Senior Policy Advisor for the Chief of Naval Operations on all Homeland Security matters. While assigned to the Office of the Chief of Naval Operations (CNO) he led a division that supported the CNO on key warfare and Homeland Security and Defense policy decisions, which included: interagency coordination, incident management, and Department of

Homeland Security/Department of Defense integration. In addition, he directed and managed the Navy Joint Doctrine program. For his work for the CNO and his efforts following the tragic events of September 11, 2001 at the Pentagon, he was awarded the Legion of Merit. As a Naval Aviator he supported operations in Iceland, Greenland, Adriatic, Mediterranean, Azores, and South America.

Mr. Breor was a Senior Executive Fellow at the John F. Kennedy School of Government, Harvard University. He received a Masters of Arts in National Security Studies and in Homeland Security and Defense from the Naval Post Graduate School, and received a Masters of Business Administration from the University of Oklahoma. In addition, he earned a Bachelor of Science in Physics from The Citadel.

Cooper, Ambassador Henry F.

Email: hcooper@ara.com

Ambassador Henry F. Cooper is Chairman of High Frontier, Chairman Emeritus of Applied Research Associates, and a Director on the Boards of The Foundation for Resilient Societies, the London Center for Policy Research, and the EMP Task Force for Homeland and National Security. He previously served as Senior Associate of the National Institute for Public Policy and Visiting Fellow at the Heritage Foundation. At High Frontier, he is working with local, state, and federal authorities to provide effective defenses against ballistic missiles, particularly those that pose an existential EMP threat to all Americans.

Since 1979, he has been appointed by the President to serve as Deputy Assistant Secretary of the Air Force with oversight responsibility for Air Force strategic and space systems (1979-81); Assistant Director of the U.S. Arms Control and Disarmament Agency, backstopping all bilateral negotiations with the Soviet Union (1983-85); Ambassador and Chief U.S. Negotiator at the Geneva Defense and Space Talks with the USSR (1985-90); and Director of the Strategic Defense Initiative (SDI, 1990-93). He served on numerous technical working groups and high level advisory boards--including the Defense Science Board, the Air Force Scientific Advisory Board, U.S. Strategic Command's Strategic Advisory Group, the Defense Nuclear Agency's Scientific Advisory Group on Effects, and the Congressional Commission to Assess the U.S. Government's Organization and Programs to Combat the Proliferation of Weapons of Mass Destruction. He received the Defense Department's Distinguished Public Service Medal, the Defense Special Weapons Agency Lifetime Achievement Award, the U.S. Missile Defense Agency's Ronald Reagan Award, the U.S. Navy Aegis BMD Pathfinder Award, and Clemson University's Distinguished Service Medal. Ambassador Cooper taught at Clemson University and worked at Bell Telephone Labs, the Air Force Weapons Lab, R&D Associates, and JAYCOR. He holds BS and MS degrees from Clemson University and a Ph.D. from New York University, all in Mechanical Engineering.

Franks, Congressman Trent

Congressman Trent Franks is a conservative, Reagan Republican, and is currently serving in his eighth term in the United States Congress. Trent serves on the House Judiciary Committee where he serves on the House Judiciary Committee, in the Subcommittee on Courts, Intellectual Property, and the Internet, and in the Subcommittee on the Constitution and Civil Justice. He is a member of the House Armed Services Committee, serving in the Emerging Threats and

Capabilities Subcommittee, and is the Vice Chair of the Strategic Forces Subcommittee. Trent Co-Chairs the Orphans and Vulnerable Children Caucus as well as the Congressional Caucus on Adoption, the Congressional Foster Youth Caucus, and the International Religious Freedom Caucus.

He is the Chairman of the National Security Working Group for the Republican Study Committee, taking the lead among the conservative members of Congress on issues relating to national security. Trent firmly believes the foremost responsibility of our federal government is to provide for our nation's common defense. He is also a staunch advocate of Israel, serving as the co-chairman of the Congressional Israel Allies Caucus.

Geary, David

Email: dgeary@poweranalytics.com

David Geary is the Vice President of Power Delivery at Power Analytics. Dave joined the company in 2015 when his company, dcFUSION, merged with Power Analytics. He is a licensed professional electrical engineer with a distinguished career spanning more than 30 years. Throughout his career Dave has brought innovation and leadership to a variety of electrical engineering projects in many markets including: government and private infrastructure engineering, design, construction, and operations. Dave is considered one of the pioneers and experts in the new evolution of higher voltage direct current (dc) power distribution for telecom, data centers, and dc microgrids—specifically, for the evolution of alternative energy integration with energy storage in dc microgrids. In his role as co-founder, principal engineer, at dcFUSION, Dave provided technical engineering innovation, vision and leadership for the use and development of alternative energy solutions, as well as higher voltage (380/400V dc) electrical system topologies for data centers and technology facilities.

Harris, William R.

Email: wm.r.harris@gmail.com

Bill Harris serves on the Board of Directors, and is the Secretary of the Foundation for Resilient Societies, a New Hampshire-based non-profit engaged in research and education on critical infrastructure protection. He is an international lawyer (Harvard, J.D., 1966) and former energy, nuclear non-proliferation, environmental, and national security project manager at the RAND Corporation. Working with physicists and engineers, he supports electric and other critical infrastructure "reliability standard" development for international, federal, and state institutions.

Jacabus, Ron

Representing Texas Senator Bob Hall

James, Dr. James J.

Executive Director, SDMMPH

John Juhasz

John Juhasz is CEO and President of Telepath Systems Inc., a scientific non-profit entity advancing the adoption of systems engineering and MBSE methods in complex systems. He is a practicing systems engineer and entrepreneur with 40 years' experience in various industry sectors, including automotive, aerospace, telecommunications and energy.

He is an INCOSE CSEP (Certified Systems Engineering Professional), and has been involved with INCOSE since its inception in 1991. He was a founder and first president of the Cleveland/Northern Ohio chapter of INCOSE, co-founded the Power & Energy Systems Working Group and the Critical Infrastructure Protection and Recovery Working Group, and chair of regional conferences focused on systems methods and energy. He is a member of InfraGard and currently serves on the Midwest Regional Board for InfraGard EMP-SIG. Mr. Juhasz has engaged in systems engineering activities in various programs in aerospace, including NASA's International Space Station and the Constellation System space exploration program, among others, developing and applying early model-based engineering methods. In automotive systems, he led OnStar system development efforts at Opel/GM Europe.

He holds ten patents dealing with innovation on vehicle information systems and adaptive braking systems. He was awarded the prestigious Rockwell "Engineer of the Year" award for his innovative work on vehicle information systems. Mr. Juhasz holds a Bachelor and Master's degree in Electrical Engineering from Cleveland State University, and a Master of Business Administration from University of Detroit. He served as Adjunct Professor at Baldwin Wallace College in the MBA program, teaching Management Decision Models and Information Systems.

Lasky, Mary

Email: mary.lasky@jhuapl.edu

Mary Lasky is a Certified Business Continuity Professional (CBCP). She has been the Program Manager for Business Continuity Planning for the Johns Hopkins University Applied Physics Laboratory (JHU/APL), and also coordinated the APL Incident Command System Team. She is the immediate Past President of the Community Emergency Response Network Inc. (CERN) in Howard County, Maryland as well as the immediate Past President of the Central Maryland Chapter of the Association of Contingency Planners (ACP). She is a member of InfraGard and the vice chair of InfraGard EMP-SIG. She is a member of the FEMA Nuclear-Radiation Communications Working Group.

She has held a variety of supervisory positions in Information Technology and in business services. For many years, she has been on the adjunct faculty of the Johns Hopkins University Whiting School of Engineering, teaching in the graduate degree program in Technical Management. Ms. Lasky is the President of the Board of Directors of Grassroots Crisis Intervention Center in Howard County, MD. She served on the Finance Committee for Leadership Howard County and is co-chair of the Steering Committee for the Leadership Premier Program. Her consulting work has included helping non-profit organizations create and implement their business continuity plans.

Manto, Charles

Email: cmanto@stop-EMP.com

Charles Manto is CEO of Instant Access Networks, LLC a consulting and R&D firm that produced independently tested solutions for EMP protected micro-grids. He received six patents in telecommunications and computer mass storage and EMP protection and assisted other entrepreneurs and investors with their intellectual property strategies. Developed valuation methodology accepted by the U.S. DOD, countries, and companies participating in industrial defense conversion.

He has facilitated due diligence of over 200 deals, managed a venture capital service, a revolving loan fund, an economic development corporation, a computer mass storage manufacturer, and broadband CLEC. Mr. Manto has also founded and leads InfraGard National's EMP SIG. He received his B.A. and M.A. from the University of Illinois at Urbana/Champaign.

Popik, Thomas (Tom)

Email: thomasp@resilientsocieties.org

Thomas Popik is chairman of the Foundation for Resilient Societies, a nonprofit group dedicated to the protection of critical infrastructure against infrequently occurring natural and manmade disasters. He is principal author of a Petition for Rulemaking submitted to the Nuclear Regulatory Commission that would require backup power sources for spent fuel pools at nuclear power plants.

Previously, as a U.S. Air Force officer, Mr. Popik investigated unattended power systems for remote military installations. Mr. Popik graduated from MIT with a B.S. in mechanical engineering and from Harvard Business School with an M.B.A.

Pry, Dr. Peter Vincent

Email: Peterpry@verizon.net

Dr. Peter Vincent Pry is Chief of Staff of the newly re-established Congressional Electromagnetic Pulse (EMP) Commission, and Executive Director of the Task Force on National and Homeland Security, a Congressional Advisory Board dedicated to achieving protection of the United States from electromagnetic pulse (EMP), Cyber Warfare, mass destruction terrorism and other threats to civilian critical infrastructures, on an accelerated basis. Dr. Pry also is Director of the United States Nuclear Strategy Forum, an advisory board to Congress on policies to counter Weapons of Mass Destruction. Foreign governments, including the United Kingdom, Israel, Canada, and Kazakhstan consult with Dr. Pry on EMP, Cyber, and other strategic threats.

Dr. Pry served on the staffs of the Congressional Commission on the Strategic Posture of the United States (2008-2009); the Commission on the New Strategic Posture of the United States (2006-2008); and the Commission to Assess the Threat to the United States from Electromagnetic Pulse (EMP) Attack (2001-2008).

Dr. Pry served as Professional Staff on the House Armed Services Committee (HASC) of the U.S. Congress, with portfolios in nuclear strategy, WMD, Russia, China, NATO, the Middle East, Intelligence, and Terrorism (1995-2001). While serving on the HASC, Dr. Pry was chief advisor to the Vice Chairman of the House Armed Services Committee and the Vice Chairman of the House Homeland Security Committee, and to the Chairman of the Terrorism Panel. Dr. Pry played a key role: running hearings in Congress that warned terrorists and rogue states could pose EMP and Cyber threats, establishing the Congressional EMP Commission, helping the Commission develop plans to protect the United States from EMP and Cyber Warfare, and working closely with senior scientists and the nation's top experts on critical infrastructures, EMP and Cyber Warfare.

Dr. Pry was an Intelligence Officer with the Central Intelligence Agency responsible for analyzing Soviet and Russian nuclear strategy, operational plans, military doctrine, threat perceptions, and developing U.S. paradigms for strategic warning (1985-1995). He also served as a Verification Analyst at the U.S. Arms Control and Disarmament Agency responsible for assessing Soviet arms control treaty compliance (1984-1985).

Dr. Pry has written numerous books on national security issues, including *Blackout Wars*; *Apocalypse Unknown: The Struggle To Protect America From An Electromagnetic Pulse Catastrophe*; *Electric Armageddon: Civil-Military Preparedness For An Electromagnetic Pulse Catastrophe*; *War Scare: Russia and America on the Nuclear Brink*; *Nuclear Wars: Exchanges and Outcomes*; *The Strategic Nuclear Balance: And Why It Matters*; and *Israel's Nuclear Arsenal*. Dr. Pry often appears on TV and radio as an expert on national security issues. The BBC made his book, *War Scare*, into a two-hour TV documentary, *Soviet War Scare 1983*, and his book *Electric Armageddon* was the basis for another TV documentary, *Electronic Armageddon*, made by the National Geographic.

Stockton, Dr. Paul N.

Email: pstockton@cloudpeak.sonecon.com

Dr. Paul N. Stockton is Managing Director of Sonecon, LLC, and directs the firm's infrastructure resilience and security-related practice. Before joining Sonecon, he served as the Assistant Secretary of Defense for Homeland Defense and Americas' Security Affairs (May 2009-January 2013). In that position, Dr. Stockton oversaw the Defense Critical Infrastructure Protection program. He led the development of the Department's *Mission Assurance Strategy* (2012), which provides analytic methodologies and policy initiatives to address DOD's dependence on civilian-owned critical infrastructure.

Dr. Stockton also served as the principal civilian advisor to the Secretary of Defense on the use of the military to support civil authorities following catastrophic events such as Superstorm Sandy, including infrastructure restoration operations. Dr. Stockton was twice awarded the Department of Defense Medal for Distinguished Public Service, DOD's highest civilian award, and received the Department of Homeland Security's Distinguished Public Service medal.

Dr. Stockton holds a Ph.D. from Harvard University and a B.A. from Dartmouth College. He is widely published in the area of critical infrastructure resilience, including: *Superstorm Sandy: Implications For Designing A Post-Cyber Attack Power Restoration System*, Johns Hopkins Applied Physics Laboratory, 2016. http://www.jhuapl.edu/ourwork/nsa/papers/PostCyberAttack.pdf

Volandt, Stephen

Email: svolandt@aurorosinc.com

Stephen Volandt is Vice President of Auroros, Inc., a contracting and management-consulting firm based in Raleigh, NC. He specializes in successfully connecting strategic purpose, risk management, decision-making, enterprise project portfolio management, operational user requirements and the technology that supports them. He has been instrumental in the establishment of governance structures for global and nation-wide organizations. Mr. Volandt co-authored the DoD CIO Executive board governance charter and participated in establishing enterprise-wide portfolio rationalization, harmonization, and transformation governance for the DoD technology portfolio.

Mr. Volandt served as the lead architect for transforming the multi-billion-dollar United States Marine Corps business enterprise to better support combat operations and readiness cycles. He provided policy, operations modeling, IT and communications modeling, planning, and budget justification for a global US Army weapon of mass destruction response capability; and was a principal operations planner and architect for joint US Army and National Guard response to smuggled nuclear weapon ground burst terrorism in the homeland. Mr. Volandt has also supported the Joint Requirements and Integration Office (JR&IO), NRO, and served as the governance team leader for the FBI's CJIS Division. Other areas of expertise include cyber business-risk assessments, technology portfolio transformation management, disaster recovery planning and operations, and military operations.

Prior to transitioning to management and technology consulting, Mr. Volandt performed or managed thousands of environmental assessments and remediation projects, to include radioactive and complexly contaminated industrial and military SUPERFUND sites, and supported state government and FEMA during several hurricane recoveries. He is a former US Marine Corps reserve officer who served at the infantry battalion, brigade, expeditionary force, and headquarters levels; specializing in readiness, decision support, exercise management, logistics, sustainment, and global operations in austere remote environments. He graduated from The Citadel with a Bachelor of Arts in Mathematics. He currently volunteers as the 2nd Vice President for the Eastern North Carolina InfraGard Chapter, as the Deputy Director, InfraGard SE Region EMP-SIG, and as the National InfraGard EMP-SIG Administrative Officer. Mr. Volandt authored the exercise scenario, exercise process, provided the maturity model for the 2015 EMP SIG annual workshop and conference. His current passion is the design, funding, and creation of resilient communities.

Woolsey, Ambassador R. James

Email: jim@odgroupllc.com

Ambassador R. James Woolsey is Chairman of the Foundation for Defense of Democracies Venture Partner with Lux Capital. He is the former Director Central Intelligence. Mr. Woolsey previously served in the U.S. Government on five different occasions, where he held Presidential appointments in two Republican and two Democratic administrations, most recently (1993-1995) as Director of Central Intelligence. From July 2002 to March 2008, Mr. Woolsey was a Vice President and officer of Booz Allen Hamilton, and then a Venture Partner with Vantage Point Venture Partners of San Bruno, California until January 2011. He was also previously a partner at the law firm of Shea & Gardner in Washington, DC, now Goodwin

Procter, where he practiced for 22 years in the fields of civil litigation, arbitration, and mediation.

During his 12 years of government service, in addition to heading the CIA and the Intelligence Community, Mr. Woolsey was: Ambassador to the Negotiation on Conventional Armed Forces in Europe (CFE), Vienna, 1989–1991; Under Secretary of the Navy, 1977–1979; and General Counsel to the U.S. Senate Committee on Armed Services, 1970–1973. He was also appointed by the President to serve on a part-time basis in Geneva, Switzerland, 1983–1986, as Delegate at Large to the U.S.–Soviet Strategic Arms Reduction Talks (START) and Nuclear and Space Arms Talks (NST). As an officer in the U.S. Army, he was an adviser on the U.S. Delegation to the Strategic Arms Limitation Talks (SALT I), Helsinki and Vienna, 1969–1970.

Mr. Woolsey serves on a range of government, corporate, and non-profit advisory boards and chairs several, including that of the Washington firm, Executive Action LLC. He serves on the National Commission on Energy Policy. He is currently Co-Chairman of the Committee on the Present Danger. He is Chairman of the Advisory Boards of the Clean Fuels Foundation and the New Uses Council, and a Trustee of the Center for Strategic & Budgetary Assessments. Previously he was Chairman of the Executive Committee of the Board of Regents of The Smithsonian Institution, and a trustee of Stanford University. He has also been a member of The National Commission on Terrorism, 1999–2000; The Commission to Assess the Ballistic Missile Threat to the U.S. (Rumsfeld Commission), 1998; The President's Commission on Federal Ethics Law Reform, 1989; The President's Blue Ribbon Commission on Defense Management (Packard Commission), 1985–1986; and The President's Commission on Strategic Forces (Scowcroft Commission), 1983.

Mr. Woolsey has served in the past as a member of boards of directors of a number of publicly and privately held companies, generally in fields related to technology and security, including Martin Marietta; British Aerospace, Inc.; Fairchild Industries; and Yurie Systems, Inc. In 2009, he was the Annenberg Distinguished Visiting Fellow at the Hoover Institution at Stanford University and in 2010–2011 he was a Senior Fellow at Yale University's Jackson Institute for Global Affairs. Mr. Woolsey was born in Tulsa, Oklahoma, and attended Tulsa public schools, graduating from Tulsa Central High School. He received his B.A. degree from Stanford University (1963, With Great Distinction, Phi Beta Kappa), an M.A. from Oxford University (Rhodes Scholar 1963–1965), and an LL.B from Yale Law School (1968, Managing Editor of the Yale Law Journal). Mr. Woolsey is a frequent contributor of articles to major publications, and from time to time gives public speeches and media interviews on the subjects of energy, foreign affairs, defense, and intelligence. He is married to Suzanne Haley Woolsey and they have three sons, Robert, Daniel, and Benjamin.

INCOSE *Insight* ARTICLES

These reprints from Insight, the official journal of the International Council for Systems Engineering (INCOSE), were first published December 2016 and are reprinted here by permission of the authors.

This issue was sponsored by the recently formed Critical Infrastructure Protection and Recovery (CIPR) Working Group, modeled substantially after the EMP Special Interest Working Group.

MODELING'S GRAND CHALLENGE–LONG TERM NATIONWIDE HIGH IMPACT THREATS

Charles (Chuck) Manto
cmanto@stop-EMP.com

Reprinted with permission of the author from the *INSIGHT* journal for December 2016, Volume 19, Issue 4, 'Critical Infrastructure and Protection.' *INSIGHT* is the magazine of International Council on Systems Engineering. It is published four times per year.

Abstract

The White House OSTP promulgated the National Weather Strategy[i] and Action Plan[ii] in April 2014 (final strategy released in Oct 2015) calling for a "whole of community" engagement for continental wide power loss for a year or more due to extreme space weather. The likelihood of such was calculated at a 6-12% likelihood in any decade. In May 2014, the DoD's Defense Threat Reduction Agency has similarly called on the small business innovation community to develop EMP protected micro grids[iii] for critical applications both on and off military bases due to the increased likelihood of long term or permanent loss of civilian power grids. Lloyds of London[iv] warned of an 18 month long black out to the corridor between DC and NYC due to extreme space weather and alerted the business community of the possible withdrawal of insurance support for those events. In 2016, the DHS published a Regional Resiliency Assessment Program for Ashburn,[v] VA and the Internet industry based there. Its key findings included the potential impact of EMP on the Internet and cyber infrastructure and encouraged industry members to hold their own seminars and exercises to examine these threats. This provides an interesting backdrop to the problem of a simultaneous loss of both power and communication grids from EMP.

On the other hand, either or both power and communications grids can be assaulted by cyber attacks which in themselves can include physical or electromagnetic attacks. For example, the recent cyber attack on the Ukrainian power grid included physical attacks. The Metcalfe substation was a physical attack on both the transformers and the communications lines of the substation.

For reasons such as these, it is prudent for government and industry to revisit their continuity plans and best practices that take a broad cross sector analysis of possible catastrophic failure modes into account. Some of the most urgent needs may be relatively simple and straightforward. Others will require a systems modeling effort to align corporate values with the business processes and technical development under their control.

Introduction:

Purpose:

The purpose for this article is to review the role of organizations such as INCOSE and InfraGard in the cross sector examination and modeling effort and apply it to the essential command and control needs expressed through electric power and communications and impact all other critical infrastructure such as water and sewer utilities.

Problem statement:

High-impact threats that can cause long-term nationwide collapse of lifeline infrastructures are complex and relatively unaddressed due to the broad cross-sector nature of the threats. How can InfraGard and INCOSE lead work to support efforts in addressing and mitigating high-impact threats to critical infrastructure that cross industry and sector boundaries?

Scope:

This article will address the difficulty of cross-sector impacts of high-impact threats to critical infrastructure and outlines collaboration already underway between INCOSE CIPR and InfraGard EMP SIG members on these issues.

Objectives:

The objectives for this article are to 1) identify key challenges presented by high-impact threats, 2) what INCOSE and InfraGard EMP SIG members may do to address them, and, 3) what next steps can be taken in the near term to make meaningful progress.

"Modeling's Grand Challenge -- Long Term Nationwide High Impact Threats"

Identifying Key Challenges

"Pre-traumatic Stress Disorder" Biases that Block the Conversation

The technical and systems complexity inherent in high-impact threats to critical infrastructure is something INCOSE members are well-suited to tackle. But, there are organizational, psychological and other biases that often shut down the discussion before it gets started. INCOSE and InfraGard members need to consider this whenever challenges of this magnitude are presented, otherwise, these biases will often overwhelm an engineer's fact-based appeals to reason. Whenever people hear something that sounds "too bad to be true", that news is dismissed. The topic can generate feelings of depression or anxiety. In a corporation, lawyers may caution staff about discussing these issues with those on the outside for fear that any unresolved vulnerability might become an invitation for a lawsuit. When working on this within the company, where does one begin, and after starting, what is the path to resources to keep at it? All of these emotionally-laden issues can contribute to something that could be called "pre-traumatic stress disorder."

Bias reinforced by older implied message of "no great threat is possible since we will rescue Americans by Day 4 of a disaster (have 3 days of food and water and all will be well)." Some Ready.gov sites continue this message while others encourage a week of preparations. While these recommendations are certainly prudent for the majority of routine emergencies, if not placed in the context of much larger though low frequency events, the need for resilience is not considered worth pursuing if a quick rescue will solve the problem, anyway.

Most if not all INCOSE members experienced the reluctance of organizations to develop systems documentation in the first place let alone keep it up to date in the second. Organizations have often succeeded in that effort by having teams developing quality or systems documentation teams assist operational teams so that operations staff members don't feel punished by having to take much time from their critical tasks to do, create or maintain systems documentation. Also, by aligning and sharing the financial benefits that accrue to the companies with the operational units that perform those systems development and documentation duties, it will be more likely that systems engineering will be done in the first place and then maintained. Providing positive reinforcement can come in the form of cash bonuses or in the provisioning of additional staff and equipment resources based on the systems documentation that justify them. Often, however, line staff and engineers who often save or make their companies significant resources wind up being punished for their efforts. Instead of seeing any benefits from the saved resources directly, those savings are often redirected elsewhere leaving those who provide the innovation and productivity with unmet needs of their own and no tangible appreciation given for their extra effort.

If these are the experiences within a given organization, imagine the greater difficulty of addressing systems issues between organizations or between industry sectors that normally do not communicate with each other.

The Widespread Simultaneous Impact of Interconnected Infrastructures

There are several factors that need to be considered in high-impact threats that are unusual. First, is the nature of high-impact threats such as electromagnetic pulse, extreme space weather, cyber attacks or coordinated physical attacks in that everything does not need to be damaged in order for the system to either collapse or cause further secondary damage. The examination and modeling of the consequences of that damage needs to include the mean-time-to-repair or the mean-time-to-replace the damaged items. Systems engineers understand the need to go even further to look at total "Operational Availability" defined to include mean logistics delay time (MLDT) and mean administration delay time. These factors include the time to order, build, and deliver given greater numbers of orders placed on possibly crippled supply chains whose production facilities may themselves need either repair or expansion in order to handle the extra orders. In some cases, the lead times may not be long, in other times, in the case of large or complex equipment, it could take over a year and longer if repair or replacement capabilities are limited and the number of items to be replaced are relatively large.

The inter-connectedness of a single impacted infrastructure needs to be fully accounted for. For example, if a physical, cyber or high-powered microwave attack on key nodes of the power grid creates a sustained regional or national power outage, then all the other critical infrastructures dependent on that power will also fail to operate even if their equipment is not damaged. In some cases, damage to one infrastructure alone, such as the power grid, could subsequently cause physical damage to other infrastructures not initially impacted, such as chemical factories that may have processes that get out of control that then create damage to the factory and to the environment. Water utilities usually are totally dependent on commercial power girds for power, yet, when that power is unavailable, water is not available to users including the other critical infrastructures most of which are severely degraded in their own capabilities once they are without water for as little as eight hours according to the July 2016 NIAC report on water infrastructure interdependencies. Coordinated attacks on multiple key infrastructures simultaneously add further complexity. An electromagnetic pulse from a high-altitude nuclear explosion is the ultimate example of a worst-case scenario because it can impact all infrastructures in a region or the continent in a single event that impacts all infrastructures simultaneously. In the case of EMP, pulses hit equipment at speeds of nano seconds and pico seconds (billionths or trillionths of seconds) with indirect cascading effects that can also happen nearly simultaneously with indirect impacts that may last hours, days or weeks as other control systems fail to act as intended or supporting systems collapse. For example, those who model these events would need to consider the loss of the power grids needed by the communications sector and the simultaneous loss of the communications sector needed by the power industry as well as the loss of water, sewer and other key resources at the same time. The lead times for restoration of all of these systems become lengthened when all these other infrastructures they need are similarly impacted.

One initial attempt at grasping such an impact was an economic impact assessment by the Sage Policy Group of an electromagnetic pulse event where industry and infrastructure was considered in the metropolitan areas between Richmond, VA and Baltimore, MD. By examining the mean-time-to-repair/replace times for just four types of systems, large communications networks, large power networks, small control systems and small electronics devices; and then comparing the loss to the economy by every economic entity by standard industrial classification (SIC) code (using the IMPLAN model (Impact Analysis for Planning) – a modeling system used to study the effects of changes across all industries in a given geographic area of the economy down to zip codes, sorted by industry looking at direct and indirect impacts.), it could be then determined what the economic losses might be depending on a low, medium or high impact event.

Assumptions on likely impacts of EMP

Infrastructure	*Percent of capacity damaged*			*Midpoint of replacement times (months)*		
	Low case	Mid case	High case	Low case	Mid case	High case
Electric grid						
Transformers	10%	40%	70%	2.5	13.5	33.0
Other	30%	40%	50%	1.5	5.0	10.0
Communications systems						
Large	10%	20%	50%	4.0	18.0	27.0
Small	5%	20%	50%	2.0	12.0	17.0
SCADA						
All types	5%	20%	50%	1.5	5.0	10.0
Electronics						
Large	20%	45%	70%	4.0	12.0	17.0
Small	1%	2%	3%	1.5	5.0	10.0
Source: Charles Manto, Instant Access Networks						

From page 14 of the 2007 Sage Policy Group, "Initial Economic Assessment of Electromagnetic Pulse (EMP) Impact upon the Baltimore-Washington-Richmond Region"[vi]

Recovery times were then computed for unprotected versus protected equipment

Infrastructure	Midpoint of replacement times with shielding (months)			Midpoint of replacement times without shielding (months)		
	Low case	Mid case	High case	Low case	Mid case	High case
Electric grid						
Transformers	0.0	1.5	27.0	1.5	13.5	33.0
Other	1.0	1.5	5.0	1.5	5.0	10.0
Communications systems						
Large	0.0	4.0	24.0	4.0	18.0	27.0
Small	0.0	4.0	15.0	4.0	12.0	17.0

Assumptions on recovery times if some strategic equipment is shielded

Infrastructure	Midpoint of replacement times with shielding (months)			Midpoint of replacement times without shielding (months)		
	Low case	Mid case	High case	Low case	Mid case	High case
Electric grid						
Transformers	0.0	1.5	27.0	1.5	13.5	33.0
Other	1.0	1.5	5.0	1.5	5.0	10.0
Communications systems						
Large	0.0	4.0	24.0	4.0	18.0	27.0
Small	0.0	4.0	15.0	4.0	12.0	17.0

The assumptions were on the generous side in that no unusual supply disruptions were considered. They were also simple in that no secondary or tertiary effects were considered. For example, if Baltimore had no functioning water delivery system, it was assumed that either there were no fires during that time. It also did not take into consideration of data disruption or controls damage that would cause an even larger event. But, despite that simplicity, it was possible to show a range of economic loss that ranged from $34B to $771B. These numbers also did not include the cost to replace damaged equipment or the economic value of the loss of life. However, the study did show that protecting approximately ten percent of the most critical infrastructure could substantially improve recovery times and minimize economic loss. Protecting those assets in place or having spares in protected nearby locations are examples of ways to accomplish that. INCOSE members could help their companies and jurisdictions look at those factors when making risk-based probabilistic choices in their mitigation efforts.

Total economic loss with and without shielding

	Post-EMP capacity with shielding			*Post-EMP capacity without shielding*		
	Low case	Middle case	High case	Low case	Middle case	High case
Total lost value of economic activity (billions)	$9	$30	$586	$34	$212	$771
Total lost value as share of annual regional gross domestic product	1.7%	5.7%	111.0%	6.5%	40.0%	147.0%
Total lost value as share of annual national gross domestic product	0.1%	0.3%	5.3%	0.3%	1.9%	7.0%

From page 23 of the 2007 Sage Policy Group, "Initial Economic Assessment of Electromagnetic Pulse (EMP) Impact upon the Baltimore-Washington-Richmond Region"

Restoration assumptions accounted for some sporadic interconnected impacts but in an admittedly simplistic way.

Comparing how unprotected versus protected infrastructure might be impacted highlights the need for modeling from INCOSE members. Whenever a given product, product line, business process or infrastructure sector is evaluated for its current vulnerability and resilience, alternative product features and business processes could be modeled to show decision makers which elements should be protected or redesigned for maximum benefit at the lowest cost. Those modeling efforts can also take a broader set of impacts into account, not unlike the Sage study or the subsequent economic studies of extreme space weather undertaken by Lloyds of London and the US Department of Homeland Security. In turn, these can be buttressed with device, product, system and industry sector evaluations of value, vulnerability, time to repair, and dependency data including the business process issues inherent with problems that impact both people and the technology they use.

INCOSE CIPR and InfraGard EMP SIG Collaboration

INCOSE established the CIPR working group to address the relatively unmet needs to understand, model, depict and then effectively communicate these cross-boundary cross-industry impacts of high-impact events. Executives within a given company or industry sector will need the assistance of INCOSE members and others who are skilled at developing system-of-systems approaches to these issued that might otherwise feel overwhelming. On one hand, INCOSE members can help simplify the depiction of these issues so that laymen in charge can have the decision support tools they need to make informed decisions. However, behind those high-level depictions, it is important to be able to drill down into the details needed to manage a sufficient number of issues including causal analysis, failure modes and effects analysis, systems and operations scenarios, diagnostic systems, life cycle support systems including real-time management allowing for the greatest possible command and control during the worst case scenarios. CIPR working group members can provide triage-like approaches that can offer near-term assistance while more complex and time-consuming approaches are developed.

It may be difficult to develop cross-industry standards and regulations, but, it is far more likely to develop resilience goals, planning tools and best practices that can address issues such as long-term regional or national power outages across multiple sectors and jurisdictions. Proposed best practices can be supported by the decision support capabilities inherent in the type of modeling done by INCOSE members. Various industry areas and infrastructure types will need machine and human interfaces to make it possible for these systems to provide real-time and near-real-time communications and support for decision makes at all levels and across all impacted areas.

An example of INCOSE CIPR collaboration with the InfraGard National EMP SIG was the development of the "Triple Threat Power Grid Exercise".[vii] This was published in 2015 by the EMP SIG (through Westphalia Press and Amazon) and covers nationwide outage scenarios of either 1, 3 or 12 month duration due either to extreme space weather, electromagnetic pulse, or cyber attack. In addition to the publication in print and Kindle form, the EMP SIG also provides read-ahead materials and facilitator guides. INCOSE CIPR or EMP SIG leaders can provide access to these other materials or discounts for case lot orders of printed material. These can be used by systems engineers and their management teams within and between enterprises to understand requirements, architecture/modeling, scenarios/operational concept documents, total life cycle support for both normal circumstances and the unwanted high impact threat disasters.

INCOSE CIPR and InfraGard EMP SIG members can provide support to local and state governments that are beginning to roll out workshops and exercises to address high-impact threats. Initial workshops have started in places such as Louisiana, Maryland, North Carolina and Texas in 2016 and are expected to be further developed in those locations and other states across the country in 2017. As local governments and industry begin to discover the difficulty and complexity of these tasks, INCOSE professionals will be useful in providing guidance and support in the development of systems approaches to these problems in each jurisdiction and their impacted companies.

A companion "Planning Guide" is expected to be published in the Fall of 2016. It will provide support for organizations and communities looking to further improve their own plans in light of high-impact threats that would overwhelm most organizations' disaster or business continuity plans. The Guide intends to facilitate cross sector collaboration and provides examples of planning processes and models that make it possible to address these threats. The first publication anticipates further collaboration and a second edition to share additional insights and progress as INCOSE CIPR and InfraGard EMP SIG continue to work on high-impact threat to critical infrastructure. This will include coverage of best-practices that will be the theme of successive editions and related publications. INCOSE members should contact their peers in the CIPR Working Group to discover ways to participate in the development and application of these resources.

References

[i] See the National Space Weather Strategy by the White House Office of Science and Technology at this website: https://www.whitehouse.gov/sites/default/files/microsites/ostp/final_nationalspaceweatherstrategy_20151028.pdf

[ii] See the National Space Weather Action Plan from October 2105 here: https://www.whitehouse.gov/sites/default/files/microsites/ostp/final_nationalspaceweatheractionplan_20151028.pdf and the October 13, 2016 Presidential Executive Order on the Space Weather Action Plan here: https://www.whitehouse.gov/the-press-office/2016/10/13/executive-order-coordinating-efforts-prepare-nation-space-weather-events

[iii] See the DTRA RFP showing vulnerability of military bases and defense critical infrastructure because of dependence on unprotected commercial power grids here: https://www.sbir.gov/sbirsearch/detail/736859

[iv] See the Lloyds of London report on EMP here: https://www.lloyds.com/~/media/lloyds/reports/emerging%20risk%20reports/solar%20storm%20risk%20to%20the%20north%20american%20electric%20grid.pdf

[v] See the DHS RRAP report on EMP effects on the cyber, data center and Internet industry here: http://coar.risc.anl.gov/ashburn-va-regional-resiliency-assessment/

[vi] See the Sage Policy Economic Assessment of EMP by Anirban Basu here: http://www.gurevich-publications.com/conspectus/economic_electromagnetic.pdf

[vii] The back cover with essential information can be copied from this Amazon page and the book or Kindle version can be ordered here: https://www.amazon.com/Triple-Threat-Power-Grid Exercise/dp/1633912493/

EVOLUTION AND RATIONALE FOR DOD'S EMP PROTECTION STANDARD

George H. Baker
Professor Emeritus, James Madison University

Abstract

The DoD EMP protection standard offers a solid basis for protecting commercial communication, data and control facilities. Because of the standard's shielded barrier and test requirements, it is not surprising that there is a strong temptation within industry and government to dismiss the MIL-STD 188-125 approach in favor of less rigorous protection methods. It is important to understand that DoD EMP protection standard for fixed facilities, MIL-STD-188-125, reflects an evolution by trial and error that spanned a period of decades beginning with the acquisition of the Minuteman Missile System in the 1960s. In fact, one of the main motivating factors for developing the standard was that system developers in the Air Force, Army, Navy and Defense Communications Agency (now Defense Information Systems Agency) and the Federal Emergency Management Agency had tried less technically-sound approaches that failed in their effectiveness, testability, and maintainability. This paper revisits the development of the DoD standard and explains its provisions and their underlying technical rationale. The paper's objective is to enable the public officials and engineers involved in planning and implementing EMP protection for critical infrastructure facilities to avoid the pitfalls encountered in the past and use the best practices available to achieve low risk protection designs that can be maintained over the entire life cycle of critical infrastructure systems.

The Department of Defense (DoD) has a long and strong experience base in protecting systems against the nuclear electromagnetic pulse (EMP). The department's EMP protection experience dates back to the 1960s with the development of the Minuteman system. In the late 1970s, recognizing the non-uniformity of EMP protection engineering practices across the services and problems with initial certification testing and life-cycle monitoring and maintenance of EMP protection, efforts were begun at the Defense Nuclear Agency (DNA) to standardize EMP environment criteria and protection engineering requirements. DNA issued the initial contract to develop a standard EMP environment waveform in 1980 to Vector Research (now Metatech, Inc.). This effort was brought to fruition by the issuance of DoD-STD-2169. In 1986 DNA, in a competitive procurement, selected SRI International to draft a protection engineering standard for fixed ground-based facilities, culminating in MIL-STD-188-125-1. In 1987, DNA assembled a technical working group to vet the standard draft material. The working group included expert government and contractor representatives from the Army, Navy, Air Force, and the Defense Communications Agency. The working group met at regular intervals from 1987 to 1991 to edit and approve the final military standard.

A major issue motivating the development of MIL-STD-188 was that, with few exceptions, the exact nature and seriousness of system malfunctions observed during EMP tests were not predicted by pre-test analysis. Analysts found that EMP effects depend on fine, often trivial, details of system construction which are difficult to model. Some details such as parasitic capacitance and inductance effects and high voltage breakdown locations are unknown, even when detailed engineering drawings were available, since these details do not influence normal system operation. Assessments based on paper studies, visual inspection, exact replication of circuit schematics, and even low-level testing were found to be unreliable.[1]

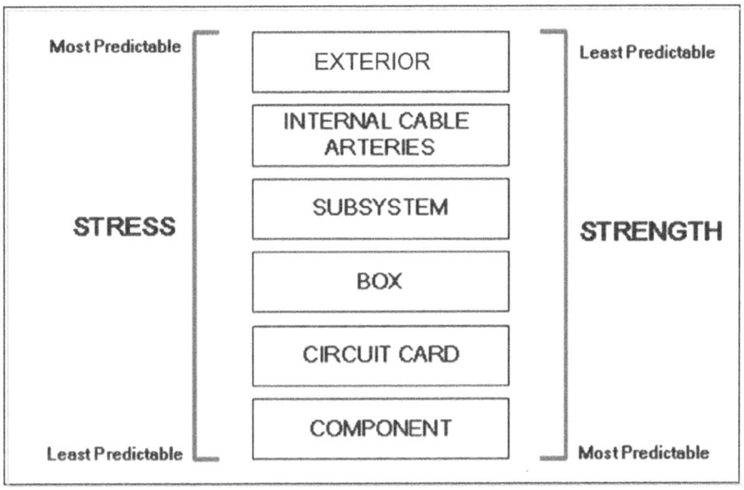

Figure 1. System EMP Coupling Layers

The assessment of system EMP vulnerability involves detailed analytical comparison of the EMP stress levels (fields, current and/or voltage levels) on the system at a selected location on the system with the strength (current and/or voltage failure thresholds) at the same location. Figure 1 illustrates the system structural layers involved in EMP system response modeling and testing. Note that uncertainties in stress levels are lowest at the system exterior (corresponding to overall system vulnerability) and uncertainties in system strength levels are lowest at the system component level. Any one of these layers could be used as the location for stress/strength comparisons. Because we are interested in overall system vulnerability, the ideal location for comparing stress with strength would be at the system exterior since we can precisely specify and impose EMP fields and penetration currents here. However, for unprotected systems, since large EMP fields and currents are allowed to flow through up to 6 coupling layers from the system exterior to internal components of the system (where actual failure occurs), it is very difficult to predict system strength to a given exterior stress. The uncertainty in system exterior strength levels is magnified over the system life cycle since internal system components are modified or replaced over time.

The difficulty of predicting systems' EMP vulnerability had a major effect on the evolution of the DoD EMP protection approach. An intense debate occurred with the EMP community from the early 1960s through the 1980s. There were two schools of thought:

1. The "tailored hardening" advocates maintained that protection should occur at the box level based on box failures observed during system testing. This approach placed a heavy emphasis on the use of terminal protection devices. The major attraction of the tailored approach was its lower up-

[1] ELECTRA Program Technical Review Group Report

front costs – i.e., systems required no initial EMP hardening and only those boxes that failed under test needed to be protected. The tailored hardening approach was also referred to as the "test-fix-test" approach.

2. The "global hardening" advocates maintained that protection should occur at the system exterior level by including an electromagnetic barrier as part of the initial system design. This approach placed a heavy emphasis on external shielding. The major attraction of the global approach was that the external barrier created an electromagnetically "quiet" interior such that unhardened commercial off-the-shelf boxes could be installed inside the system. The downside was higher initial costs associated with shielding materials. The global hardening approach was also referred to as the "fix-test-fix" approach.

The impasse precipitated two high-level reviews. The first review conducted by a special convocation of the Defense Science Board (DSB) during 1979-80 focused on aircraft protection[2]. The second, conducted by the National Research Council of the National Academies of Science during 1982-84 looked at EMP protection of systems in general.[3]

The Defense Science Board Task Force Report addressed three protection approaches: (1) global system shielding, (2) adaptations involving protecting only mission-essential equipment, and (3) component hardening. They identified two major sources of uncertainty underlying their recommendations - i.e., uncertainties in EMP coupling level (stress) analysis, and uncertainties in overall system failure prediction (strength) analysis. The DSB concluded that the tailored protection approach had an excessive reliance on analysis, and the associated uncertainties in stress and strength levels did not provide high-confidence in system survivability. The Board recommended that the best approach to EMP protection is to use a minimum number of contiguous shields with the smallest possible number of EMP penetrations (i.e. penetrating wires, doors, pipes, and apertures). They acknowledged that adaptations involving isolating mission critical equipment would permit cost-savings by reducing necessary shielded volumes.

It is noteworthy that the National Research Council (NRC) committee came to the same conclusions as the DSB. Like the DSB, the NRC report emphasized the large uncertainties inherent in the entire process of estimation of currents and voltages at the component level and the associated protection requirements. Their solution was to recommend controlling system stress to well-known values by using integral shielding and penetration control at the system exterior. They also recommended great emphasis on developing better and less expensive means for virtually complete and effective shielding of systems. Their idea was to put up a tight shielded barrier to keep the energy out of the system. This creates an interior "quiet zone" such that there is no need to predict what happens inside thus avoiding the associated prediction uncertainties. With the proper engineering, barriers can be designed with levels so low that virtually any type of electronic component will survive. The best way to reduce electronics' strength uncertainties is to provide a system design that limits EMP fields and currents at the system's exterior to levels known to be generally safe – i.e., levels comparable to normal signal background noise.

The National Research Council committee also recommended the development of design strategies that are testable. A major advantage of global shielding is that testing is much easier – one must validate only that the outer shield is effective. The NRC report stated that if the system is shielded so well that virtually

---

[2] Report of the Defense Science Board Task Force on EMP Hardening of Aircraft, Office of the Secretary of Defense, 1980

[3] National Research Council, Evaluation of Methodologies for Estimating Vulnerability to Electromagnetic Pulse Effects, National Academy Press, Washington, D.C. 1984

nothing gets through, periodic tests of the integrity of the shield and high-level current injection tests on all cables entering the shielded enclosure constitute an adequate test. This is important not just for initial system acceptance tests, but for life cycle hardness surveillance testing required for EMP hardness maintenance. If the outer system shield is certified effective, interior electronics can be swapped and replaced over the system life cycle without fear of compromising system EMP survivability.

The NRC was critical of the tailored hardening approach stating, "The Committee is uncomfortable with the use of tailored hardening to design new systems." They observed that the component-level protection rather than the system-level protection carries much more risk of vulnerability, again pointing to inaccuracy of analytical EMP coupling and system response predictions. It is straightforward to specify or predict EMP levels outside a system, but as EMP energy propagates inside the system, internal wiring complexity makes it very difficult to predict the levels of currents and voltages that flow to individual components. The crux of the global versus tailored protection issue was that this is where the tailored hardening proponents were attempting to protect. It should be noted that the NRC did not entirely dismiss box level protection, but observed that protection concepts involving shielding at the box level could be admissible provided that optical fiber was used for all box interconnections.

The global shielding approach used by DoD beginning in the 1980s follows the recommendations of the Defense Science Board and National Research Council committees. MIL-STD-188-125-1 embodies the global shielding approach for fixed ground-based communication and data centers. MIL-STD-188-125-2 embodies the global shielding approach for portable and ground-mobile systems.

Systems that are electromagnetically simple with a contiguous shield and having a small number of protected penetrations have been engineered by DoD and demonstrated to be survivable to EMP. The engineering principles are straightforward as illustrated in Figure 2:

1) Make the electronics portion of the system as compact as possible.
2) Enclose these electronics in a single continuous shield.
3) Limit the number of electromagnetic penetrations through the shield and protect all remaining penetrations.
4) Certify the hardness of the system via shielding effectiveness tests and current injection tests of cable penetrations.
5) Periodically retest the shielding and penetration protection integrity to maintain hardness over the system life cycle.

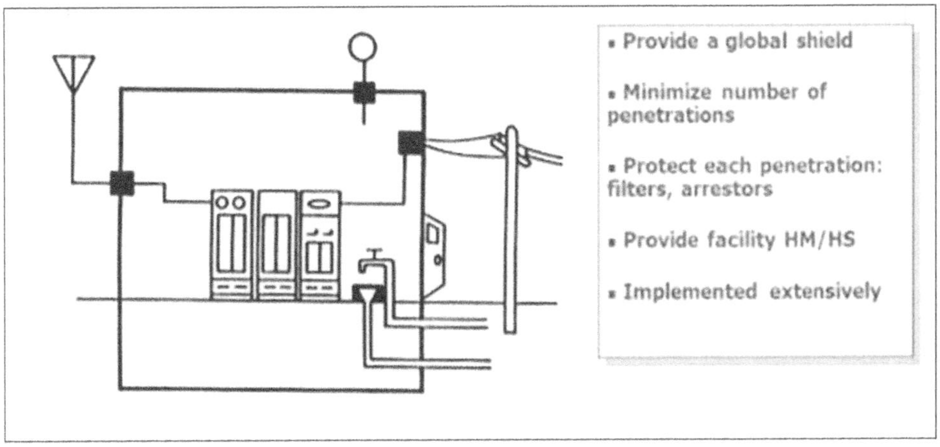

Figure 2. Notional Illustration of the Global Shielding Approach for Fixed Facilities

Numerous systems have successfully implemented this approach in their engineering design and fabrication. A partial list includes:

Missiles (e.g., Minuteman Upgrade, Peacekeeper, Air Launched Cruise Missile);
Aircraft (e.g., B1-B, TACAMO, E-3A, E-4B); and
Ground Systems (e.g., Ground Launched Cruise Missile, MILSTAR, Ground Wave Emergency Network, National Military Command Center, Alternate National Military Command Center, Defense Satellite Communication System (DSCS) Ground Terminals, Pershing launch control system)

An illustration of the MIL-STD-188 facility outer boundary shield implemented on the DSCS ground terminals is included in Figure 3.

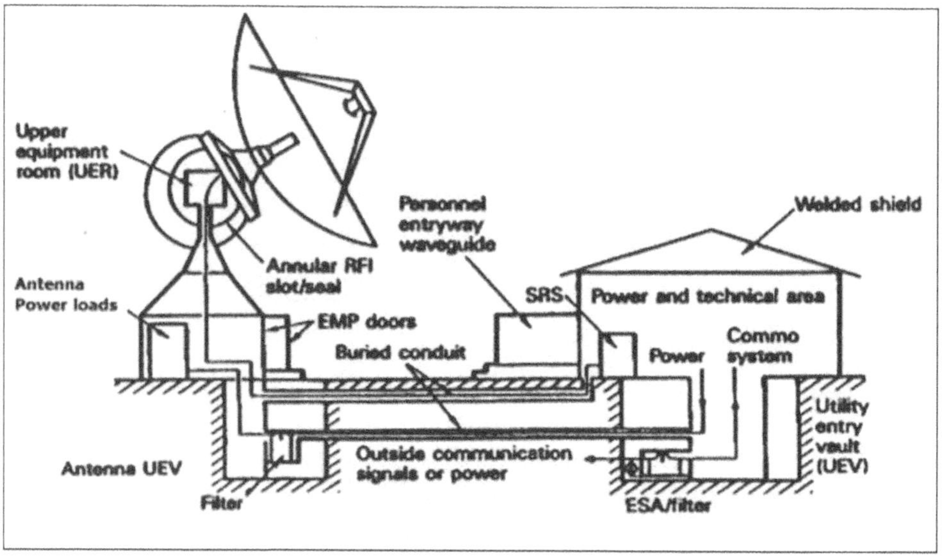

Figure 3. MIL-STD-188-125 Protection Design Example – DSCS Ground Terminal

It is important to note that the DoD standards are test-performance based. Different shield designs and penetration protection devices may be used, as long as the shielding effectiveness and penetration currents meet the specified final acceptance test.

When compared to the cost of lost societal services, EMP protection costs for critical infrastructure systems are well justified. For long-term outages of the electric power grid caused by EMP, economic losses would be measured in trillions of dollars.[4] The costs to harden are a small fraction of system cost for new systems. For example, DoD has found that EMP protection costs for large buildings amount to 3 to 8 percent of total system cost if included in the initial system design.[5] And costs are manageable for retrofit protection of existing systems by isolating critical electronics within shielded cabinets or rooms.

[4] G.H. Baker, Cost-Effective EMP Protection for Communications Networks and Power Sources, Proceedings of 2013 Dupont Summit on Mitigating High-Impact Threats to Critical Infrastructure
[5] Frank Gertcher, Benefits and Costs of Protecting Infrastructure Systems Against Terrorist and Related Threats: Cost Analysis, Defense Threat Reduction Agency, Report No: RT-0103-99.

It is noteworthy that EMP protection has many benefits in addition to EMP immunity. The MIL-STD-188-125 EMP protection design enhances signal emanation security (TEMPEST protection). In addition, DoD standard protection reduces RF weapon effects, and is effective against solar geomagnetic disturbance effects, electric power outages caused by line transients, lightning effects, and electromagnetic interference.

WILL SYSTEM-OF-SYSTEM MODELS SUPPORT COST RECOVERY FOR CRITICAL INFRASTRUCTURE RESILIENCY?

William R. Harris, williamh@resilientsocieties.org

ABSTRACT

The fragility of critical infrastructures and the inter-dependency of infrastructures such as electric grids, water, wastewater, agriculture and manufacturing supply chains should encourage cost recovery for investments in *resilience,* not merely system *reliability*. Most cost recovery standards support investments for safety and reliability, but not *resilience.* If model based system engineering (MBSE) and virtually all other engineering domains can demonstrate greater system protection, adaptability, and recovery (minimizing cascading harms to connected critical infrastructures), regulatory standards should provide cost recovery for investments in *resilience.*

Beyond cost-recovery incentives, system-of-system engineers can design facilities for supply component availability, surge production and other commitments to more robust supply chains. Firms that plan for supply chain disruptions can expand their market shares during and after emergencies, while other firms decline or fail entirely. System designs for greater resiliency may reward firms that overcome system shocks, even when resiliency investments are not automatically priced into goods sold or services delivered.

KEYWORDS

Cost-recovery, critical infrastructure, resilience

ADVANCING FROM SYSTEM RELIABILITY TO INTER-SYSTEM RESILIENCY

In recent decades, systems and system-of-systems (SoS) engineers have advanced the goal of developing critical infrastructures that can withstand shocks, that can adapt, and even self-heal. (Amin, 2002; Amin and Giacomoni, 2012). A purpose of this article is to highlight regulatory opportunities for the pass through of costs to taxpayers, infrastructure ratepayers, or market purchasers, a class of *cost recovery*, if Model Based Systems Engineering (MBSE) architectures demonstrate that investments will be more cost-effective if designed to increase infrastructure *resilience* than to increase infrastructure *reliability*. What is the difference?

Reliability involves the probability of successful mission performance. *Resilience* involves the ability to protect, to absorb shocks, to adapt and to recover more quickly, more economically, or more smartly, as by prioritizing the delivery of goods or services to interdependent infrastructures so they avert cascading service outages. A program of *resilience* should generally enable protection of life and property at lower costs than investments in system *reliability* alone.

How can system of system engineers demonstrate that investments in *resilience* deserve at least equal opportunity for cost-recovery as investments in system reliability?

CRITICAL INFRASTRUCTURE: RELIABILITY AND COST RECOVERY

In the United States, most regulated industries, such as the bulk power system or electric distribution systems regulated by states authorize cost-recovery from ratepayers for initiatives that increase system *reliability* but not system *resilience*. Under the Energy Policy Act of 2005, a system of *reliability standards* is authorized by the Federal Energy Regulatory Commission (FERC) "to provide for reliable operation of the bulk-power system." [16 U.S. Code § 824o(a)(3)]. Electric utilities that improve the reliability of the regional high voltage transmission system are eligible to pass along regulator-approved reliability upgrades to wholesale and, for the states, to retail ratepayers. The states retain, under "savings provisions" the authority "to take action to ensure the safety, adequacy, and reliability of electric service within that state…." [16 U.S. Code § 824o(i)(3)].

However, for upgrades to electric generators, under FERC orders, most upgrades to system reliability are costs absorbed by the generator plant owners, who compete on price when selling blocks of power within regional markets. This is one of the consequences of market deregulation, which has generally resulted in lower costs for consumers but not greater system resilience. In fact, transformation of the electric industry to meet environmental goals has resulted in higher dependency upon natural gas generation (in the U.S. about 33% and rising), and reduced coal-fired generation (about 32% and declining). From a *resiliency* perspective, replacing on-site fuel supplies with long distance gas pipelines that are shifting to all-electric pipeline compressors increases net risks of cascading electric outages.

There are exceptions to assure that high priority customers, such as nuclear power plants, will be able to compensate other electric generator owners for reserving generator capacity, fuel and "cranking paths" to assure *system restoration* in emergencies. This is known as *black starting*, and FERC allows cost recovery under the Federal Power Act for these vital services.

What would be the appropriate mechanism for cost recovery if a nation sought to protect transformers from the geomagnetic induced currents (GICs) that cause thermal overheating of transformers in a solar storm? In September 2016, in FERC Order 830, the Commission established a process by which utilities can model their solar storm vulnerabilities and apply for cost recovery to block the harmful currents that risk loss of transformers in solar storms. There are neutral blockers that have been third-party tested that are available for purchase. In the United Kingdom, electric companies can also apply for and receive cost recovery for investing in extra spare transformers to improve reliability and to prevent long term blackouts following solar storms.

For water and wastewater systems, there is no comparable federal system of cost recovery. A municipal water utility may obtain the approval of a city-appointed water board to increase water rates to attain higher service reliability or to assure the safety of water quality, also meeting federal standards for water quality and safety. Other municipalities have sold their water systems, and contract for water and wastewater services with for-profit enterprises.

The overall result is that most water systems in the United States are not designed for and operated to provide *resiliency* for essential water and wastewater operability if for whatever reasons the regional electric grid cannot provide continuous electrical service or recover from a grid blackout within a matter of hours.

A recent report from the National Infrastructure Advisory Council (NIAC) provides an overview of the dependencies of the various critical infrastructures upon water. Table ES-1 from that Report shows the rapid projected decline of service delivery by other critical facilities if electric power is not restored to assure water services within hours, rather than days. If time-to-recovery of water systems, or expected loss of life were part of both intra-*sector* and *inter-sector resiliency modeling* the same rate of system degradation might remain but the likelihood of extended inoperability of water services could decline. Or, backup water storage and backup wells with renewable energy support might proliferate in support of inter-sector resilience.

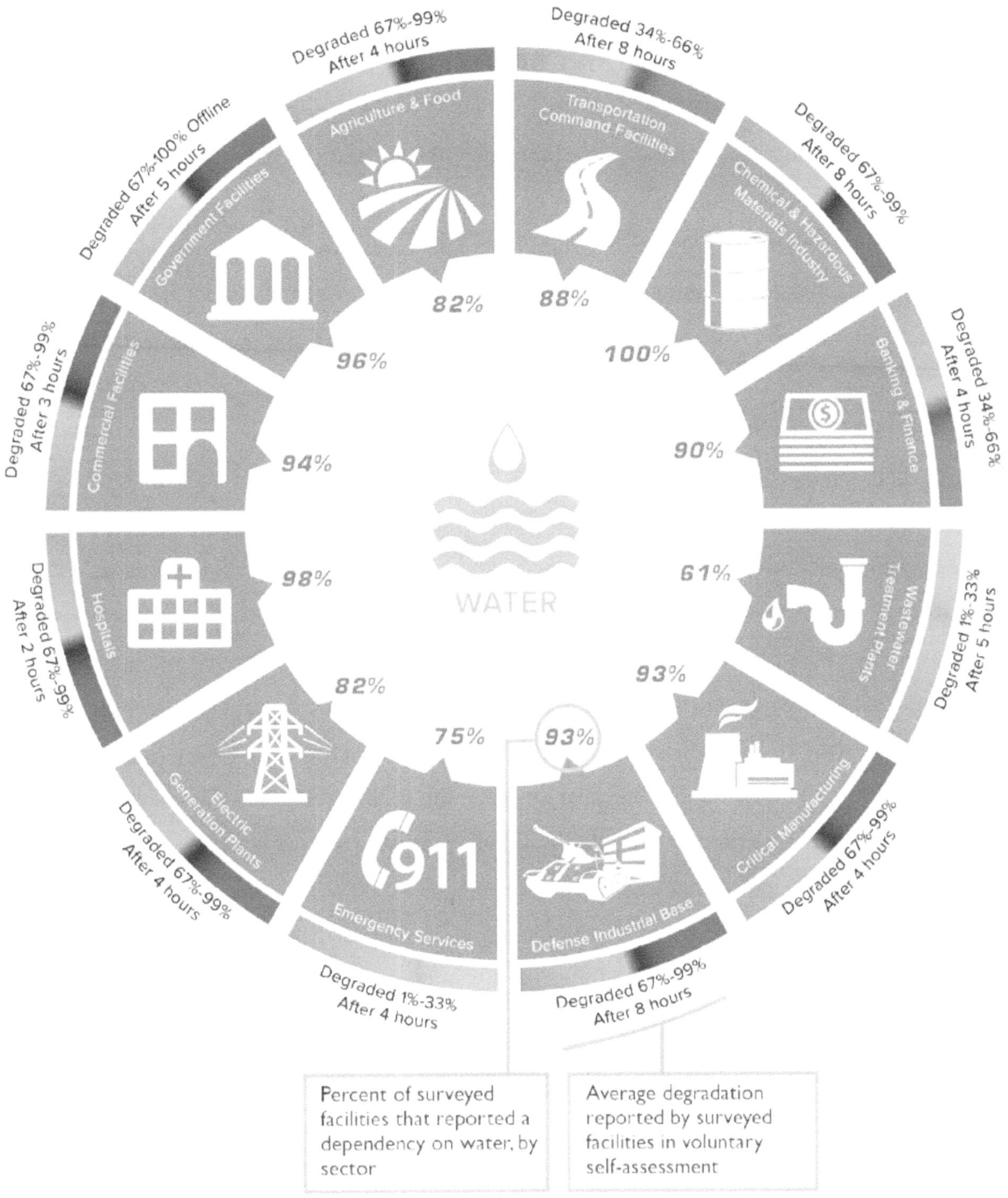

Figure 1: National Infrastructure Advisory Council

Water Sector Resilience: Final Report and Recommendations, June 2016

CAN INCOSE SYSTEM ENGINEERS MODEL INTER-SECTOR RESILIENCY AND ENCOURAGE REFORM OF COST RECOVERY POLICIES?

INCOSE system engineers might consider the potentially significant benefits of developing system architectures for existing critical infrastructures; their interdependencies on other critical infrastructures; single or common points of failure; and future redesigns that strengthen self-diagnosing, adaptive, and partially self-healing systems. Understanding the *as-built system interdependencies* may be a pre-condition of significant improvements. This is particularly a concern for owner-operators, because the burden of proof remains with those owners when they apply for cost recoveries that improve either *reliability* or *resilience.*

System engineers, who might prefer to model optimal system designs for critical infrastructures, must also address the current state of infrastructure requirements versus more limited current capabilities to meet existing and planned *safety* and *reliability standards.* These standards may create barriers to "best practices" for resiliency because they neither guarantee cost recovery beyond minimal reliability or safety standards, nor do they include as quantifiable benefits in a rate-making case designs for faster recovery times for other critical infrastructures.

For example: if an electric grid is designed to promptly separate islands for electric generation supporting water and wastewater systems, the traditional rate-making standards will not recognize the benefits, during grid blackouts, of faster recovery of water or sewer services because of "resilient island" design enhancements.

In the United States almost all of the standards for federal and state regulation of critical infrastructures enable cost recovery for *reliability* but not for *resiliency. Why is this a challenge?* Because many of the "best buys" for resiliency, including adaptability and expedited or cost-effective recovery capabilities, are not guaranteed an allowable cost recovery within either regulated or deregulated markets.

BENEFITS OF "ALL HAZARDS" MODELING

Over the past year, volunteers for a committee within the InfraGard National Members Alliance ("InfraGard"), a U.S. non-profit, have addressed a broad range of man-made and natural occurring hazards to electric power grids. The Electromagnetic Pulse Special Interest Group of InfraGard examined five threats to critical infrastructures. Their book analyzed critical infrastructures in North America; a comparable review for other continents would find parallel vulnerabilities to: high-altitude electromagnetic pulse (HEMP) weapons, severe solar geomagnetic storms, cyber attacks, physical attacks, or Radio Frequency (RF) attacks on critical infrastructures.

In the resulting book, *Powering Through: From Fragile Infrastructures to Community Resilience* (Electromagnetic Pulse Special Interest Group, 2016), a team of 25 volunteers analyzed the consequences for loss or degradation of critical infrastructure. Of special concern were identified intra-sector fragilities, such as the joint dependencies of the agricultural sector for electricity, water, and chemicals. Of greater concern are "single points of failure" that have not been systematically modeled.

What are the consequences of policies that seek to upgrade infrastructure reliability, addressing single hazards rather than performing "all hazards" analysis of *resilience* and *recovery*? Consider the ongoing efforts in the United States to protect the electric grids from naturally occurring solar-caused electromagnetic pulse (EMP) hazards, but not from man-made high altitude electromagnetic pulses, known as HEMP. Both varieties of EMP induce slow current surges (lasting for minutes) on long electric lines, known as E3 pulses. But the man-made EMP weapons also produce far faster high voltage pulses (lasting for a fraction of a microsecond), known as E1 pulses. These E1 pulses do not occur during solar geomagnetic disturbances. But if electric grid planners neglect to model the E1 vulnerabilities, a wide range of E1-vulnerable electronic systems will risk damage to both communication and water systems needed for the electric grid to function.

Furthermore, the peak currents produced by the man-made E3 are factors of several higher than natural, solar-caused E3. So a blocking device adequate to protect transformers from solar storms would risk transformer losses in a HEMP attack. Without considering both hazards together, the electric utility industry might seek approvals and cost recovery for only the solar storm E3 hazards. For only minimal additional costs, neutral blockers that would protect transformers from solar storms could also protect against the more intense E3 currents resulting from man-made HEMP events.

Moreover, because solar storm E3 has a propensity to damage electrical equipment at more northerly latitudes or in coastal regions, if the transformers eligible for cost-recoverable protective equipment are all in the northern tier states plus some coastal states, the nation's three regional interconnection systems would be remain vulnerable to prolonged electric blackouts from man-made EMP attack. For that hazard, the lack of protection against E3 pulses throughout most of the electric grid would render impractical separate equipment to protect against E1 pulses, risking loss of thousands of transformers and multi-month or multi-year electric blackouts.

The Earth's magnetic field orientation causes the greatest intensity of E3 and E1 pulses southward of the geospatial locus of detonation. So an integrated mitigation program is not feasible if each hazard is analyzed as if it is unconnected to related hazards.

There is no substitute for "all hazards" systems engineering modeling, analysis and design if cost-effective *resiliency* is a public policy goal. Some of the same control facilities benefitting from electromagnetic pulse protection could also benefit from concurrent cyber security protections. Both modeling and applying holistic system thinking for all three hazards concurrently would make sense, and might encourage policies supporting cost recovery for greater *resilience* and not merely single hazard system *reliability.*

A PRECEDENT FOR COST-SHARING: SMART GRID INVESTMENTS IN 2010-2015

Even if "all hazards" modeling of critical infrastructures is feasible, systems engineers (and particularly SoS engineers) need to recognize that the regulatory systems often create financial barriers to protection. Without this knowledge, system-of-systems resiliency models may lie on the shelf and not be implemented.

For example, the Foundation for Resilient Societies is currently modeling protection of the Texas electric grid from both high altitude EMP (HEMP) hazards and from solar storms (GMD). Preliminary cost estimates for protection of most or all of the Texas electric grid show that protection of generating facilities is substantially more than half of the total projected grid protection costs. System-of-system engineers need to be able to model the major components of system protection costs, to know if there may be a cost-recovery barrier.

Generator owners are usually expected to pay for their own system improvements, and compete on wholesale contract prices within regional transmission markets. So it is not surprising that generator owners have not embraced man-made EMP protection as a mandate for the electric utility industry.

Is there a cost-sharing solution?

Under the American Recovery and Reinvestment Act (ARRA) legislation in year 2009, the Congress authorized federal taxpayer cost-sharing for "smart grid" equipment installations, so long as electric utilities paid at least a 50% match for these taxpayer expenditures. The federal government paid about $4.5 billion to accelerate the "smart grid" which has the benefit of enabling emergency demand reductions and load balancing, and the concurrent hazard of extending the *cyber-attack* surfaces to facilities with potential cascading harms. The private sector did more than match the federal grants, and expended over $8 billion of their funds in smart grid technology. (DOE, 2014; DOE, 2015).

This "Smart Grid" program of 2010-2015 did not include cybersecurity requirements. Many utilities obtained cost-sharing to install smart meters and synchrophasors but without installing encryption. Subsequently, following the remote manipulation of industrial control systems within three Ukrainian electric distribution systems in December 2015 caused a blackout affecting about 230,000 customers. Concerns about vulnerabilities of other electric grids resulted. (Khan, et al. 2016). FERC has ordered additional cyber-protections for control centers duties (Order 822, Jan. 2016) and future protection of cyber supply chains (Order 829, July 2016).

If in a future session the U.S. Congress were to offer a 50 percent matching eligibility for EMP protection investments – both solar and man-made hazards -- as has been done for the "smart grid" in years 2010-2015, a condition of participation could also include mandatory protection against cyber-physical attacks. With taxpayer co-funding, generator owners would have a more powerful incentive to participate, and transmission owners could recover lesser additional costs of transmission system hardening through electric ratepayers.

There is also a factor of (approximately) ten cost savings inherent in including EMP protection in the system *design phase* rather than as a *retrofit*. System-of-system modeling with domain systems engineering expertise could illuminate and calculate component costs of protection, as well as model the benefits of alternative packages of protection, mitigation, and cost-sharing in a private-public partnership.

SUMMARY AND CONCLUSIONS

We face prospects for *market failures* because cost recovery varies by type of equipment or service, and the legal system rewards system *reliability* but not system *resiliency.* With better modeling, some reliability improvements may be demonstrated to reduce *resiliency*. System modelers need to provide display capabilities understandable by lawyers, economists, and legislators. Because resiliency is a better buy than mere reliability, and because modeling by itself is no guarantee of resilient outcomes.

A recent paper on cyber-physical resilience (Arghandeh, Meier, Mehrmanesh and Mili, 2016) explains that *resilience* requires timely action, not just *risk assessment:*

> "A resilient system must go beyond risk assessment and carry out a set of actions in a timely manner to ensure adequate system functionality in the face of risks, sudden changes and threats… We posit that the concept of resilience in infrastructural networks must be centered on more holistic, rigorous and temporal analyses than those typically performed in traditional risk assessment."

Model-based system engineers have a challenge: to demonstrate that *resilience* achieves greater societal protections and global productivity than incremental investments in system *reliability.* Resiliency engineers can demonstrate that "all hazards" modeling and cross-infrastructure investments can accelerate recovery of global infrastructures and supply chains upon which we all depend. Cost recovery for *resiliency* will follow better modeling.

REFERENCES

Amin, S. Massoud 2002. "Toward secure and resilient interdependent infrastructures." *Journal of Infrastructure Systems* 8(3), 67-75.

Amin, S. Massoud, and Anthony M. Giacomoni 2012. "Smart grid -- Safe, secure, self-healing." *IEEE Power Energy Mag.* 10(1), 33-40

Arghandeh, R., A. von Meier, L. Mehrmanesh, and L. Mili, 2016. "On the definition of cyber-physical resilience in power systems." *Renewable and Sustainable Energy Review* 58, 1060-1069. https://arxiv.org/ftp/arxiv/papers/1504/1504.05916.pdf

Electromagnetic Pulse Special Interest Group of InfraGard, 2016. *Powering Through: From Fragile Infrastructures to Community Resilience (Version 1.0* (forthcoming).

Khan, R., P. Maynard, K. McLaughlin, D. Laverty & S. Sezer (2016). "Threat Analysis of BlackEnergy Malware for Synchrophasor based Real-time Control and Monitoring in Smart Grid." *4th Int'l Symposium ICS & SCADA Cyber Security Research. BCS.* pp. 53-63, at

.http://ewic.bcs.org/upload/pdf/ewic_icscsr2016_paper7.pdf
National Infrastructure Advisory Council, *Water Sector Resilience Final Report and Recommendations*, June2016:
https://www.dhs.gov/sites/default/files/publications/niac-water-resilience-final-report-508.pdf
U.S. Department of Energy, Office of Electricity Delivery & Energy Reliability. 2014, *The American Recovery and Reinvestment Act highlights: Jumpstarting a Modern Grid.* http://energy.gov/sites/prod/files/2014/12/f19/SGIG-SGDP-Highlights-October2014.pdf
U.S. Department of Energy, Office of Electricity Delivery & Energy Reliability. 2015. *ARRA Grid Modernization Investment Highlights – Fact Sheet.* http://energy.gov/oe/downloads/arra-grid-modernization-investment-highlights-fact-sheet.

ABOUT THE AUTHOR

William R. Harris is an international lawyer. He obtained a B.A., Harvard College, 1962, and J.D. Harvard Law School, 1966. At the RAND Corporation he developed computer-based expert systems to assure constitutional continuity of government and recovery assistance to the executive, legislative and judicial branches of federal and state government organizations. He also researched Hot Line modernizations, designed Nuclear Risk Reduction Centers, and participated in energy and environmental program research for state and federal sponsors. He serves as a Board Member and principal attorney for the Foundation for Resilient Societies, participating in electric reliability and cybersecurity standard rulemakings of federal regulatory commissions.

Modeling Emergency Planning and Recovery

Michael Campbell, michael.campbell@specinnovations.com
Steven H. Dam, steven.dam@specinnovations.com

Abstract
The purpose of this paper is to establish the benefit of planning using model-based systems engineering (MBSE) techniques prior to a hazard that could affect critical assets and to identify the risks associated after an event has occurred to establish the means of recovery for critical assets. MBSE provides a systematic approach to developing executable plans through visualizing the processes; identifying resource limitations and gaps; identifying investment requirements for mitigating and recovery from disasters; and estimating the costs associated with the entire emergency planning, preparedness, and recovery.

The scope of this paper includes the policies that have been set forth by domestic and international government and non-governmental organizations that cover the methods of planning and recovering from an event. We will also discuss the benefits of modeling and how to utilize systems engineering methods and tools to allocate resources, identify gaps, itemize cost schedules, develop timelines, and use simulation to aid in the planning process. The proper use of technology and protection of networks during an event allows for the planning process to transition smartly into recovery.

The background for this paper comes from applying MBSE to disaster management and recovery for the US Government, which resulted in the development of the Assessment of Disaster Readiness Tool (ADRT). ADRT is a strategic concept and framework, which includes a measurable baseline assessment of a designated partner nation's disaster readiness / resiliency level, supported by a strategic Disaster Management (Response) Plan (DMP/DRP) for improvement. ADRT is free and can be used by planners worldwide.

Introduction

Emergency planning and recovery are important to any infrastructure design. Everyone recognizes the importance of planning prior to a hazard that could affect critical assets. Part of any planning approach can include, but is not limited to, the following: identifying the potential risks; assessing the damage; and recovering the capabilities of critical assets. However, many of these plans are developed without using model-based systems engineering (MBSE) techniques and can result in a non-executable plan (i.e., one that will not work in unpredicted situations).

Typically, a person or group of people author the document from templates, previous works and lessons learned. The plan is then limited by the group who built its experience and vision for the potential disaster. Many assumptions are made, but few are usually documented. Little time and energy is spent creating a good set of scenarios and modeling those scenarios to determine if they properly represent the breadth and depth of the potential impacts from a variety of disasters.

In addition to creating a good set of scenarios, analysis of policies and standards in Disaster Management must occur prior to detailed planning to ensure that the requirements derived from those policies are met. These policies and standards represent the best practices for the community and enable support between states and countries during a time of crisis.

The Assessment of Disaster Readiness Tool (ADRT) is a free tool that utilizes MBSE techniques and can be used by planners for emergency planning and recovery. The systems engineering process that we used for developing t ADRT followed the standard requirements analysis, functional analysis, and synthesis steps originally developed in draft MIL-STD-499B. *Figure 1*[i] shows our adaption of this standard. The next sections discuss elements of this process and how we used them to develop the tool.

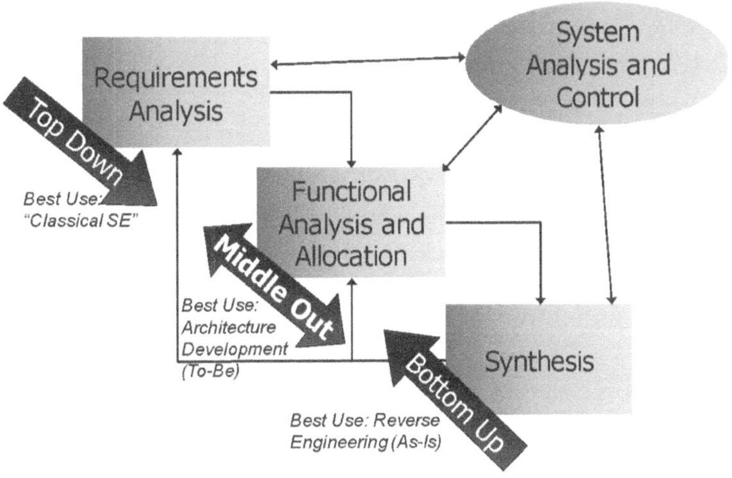

Figure 1. SPEC Innovations' Standard Design and Analysis Process

Requirements Analysis:
Analyzing the Key International Policies Affecting Disaster Management

The requirements analysis process we used begins with gathering requirements wherever they occur. In the area of disaster management, a number of policies have been developed by the international community resulting in "frameworks" that provide strategic goals, objectives, and, in many cases, concrete steps for the community. We discuss a few of these below.

Twenty years ago, the United Nations (UN) held the Yokohama conference[ii] in Yokohama, Japan. The concern of this conference was the effect natural disasters were having on human lives. Ultimately, the State Nations of the UN affirmed:

1. Disaster prevention, mitigation, preparedness and relief are four elements that contribute to and gain from the implementation of sustainable development policies.
2. Disaster prevention, mitigation and preparedness are better than disaster response in achieving the goals and objectives of the Decade. Disaster response alone is not sufficient, as it yields only temporary results at a very high cost.
3. The information, knowledge and some of the technology necessary to reduce the effects of natural disasters can be available in many cases at low cost and should be applied.
4. The adopted Yokohama Strategy and related Plan of Action for the rest of the Decade and beyond will develop and strengthen national capacities and capabilities and, where appropriate, national legislation for natural and other disaster prevention, mitigation and preparedness, including the mobilization of non-governmental organization and participation of local communities.

The Hyogo Framework[iii] followed Yokohama in 2005 and again set high standards for Disaster Management for the international community. The expected outcome for the Hyogo conference was much the same of the Yokohama conference:

The substantial reduction of disaster losses, in lives and in the social, economic and environmental assets of communities and countries.

After Hyogo came the Sendai conference.[iv] This conference still had the same expected outcome as stated above, yet the Sendai framework was extended beyond a decade for desired results. The Sendai conference met in 2015 and extended their international plan to 2030. The main reason for 25 years of expectations in Disaster Risk Reduction is that the severity of disasters has always undermined any infrastructure efforts internationally in underdeveloped countries. The same could be said for the United States.

In addition to the frameworks above, we reviewed a number of others, including the Failed States Framework,[v] ICAF,[vi] OCHA,[vii] and MPICE.[viii] In each of these frameworks, we found the basis for potential requirements of disaster planning and recovery systems. The goals and objectives of these frameworks were captured in a MBSE tool and then analyzed to create requirements statements. However, after analyzing these requirements, we were concerned that we may have missed key functional requirements. As a result, we decided that we needed to validate the requirements through functional analysis.

Functional Analysis: Developing Planning Scenarios

Classic functional analysis usually starts with higher-level functional requirements, such as the ones we derived from the frameworks above. Then a functional model is created from these requirements. However, another approach is to use scenarios to elicit the requirements from stakeholders using examples of realistic disasters, whether natural or man-made.

Using MBSE, we developed scenarios that enabled Emergency Planners to model and simulate their plans with cost, schedules, and resources to identify gaps in the process. Planners can also conduct assessments of their own critical infrastructure to determine how vulnerable they are to various threats and hazards. The key is to develop a reasonable set of scenarios that cover the breadth of the trade space. *Table 1* below shows an example of such a set of scenarios. Note that the characteristics of each scenario were varied to ensure that the breadth of the scenarios covered all the critical factors. For example, the Classes (third column of *Table 1*) represent the different categories of States - i.e., Class I is Fully Functioning, Class II is Partially Functioning, and Class III is "Failed" or Failing. An example of a Class III State is modern Syria.

Table 1. Scenarios Used to Develop Data-Gathering Questions for ADRT Development

Scenario	Permissiveness	Class	Disaster Type	Disaster Scope	Acceptance of Assistance	Disaster Management
Scenario 1: Low category hurricane strikes Readiness Level I Partner Nation, no war	Permissive	I	Predictable	Localized	Willing	Able
Scenario 2: Earthquake magnitude of 6.0+, in a Readiness Level I Partner Nation with, no war	Permissive	I	Unpredictable	Localized	Willing	Able
Scenario 3: Medium category hurricane strikes a Readiness Level II Partner Nation amidst civil unrest	Permissive	II	Predictable	Localized	Willing	Able
Scenario 4: Tsunami strikes a Readiness Level II Partner Nation during a civil war	Permissive	II	Predictable (limited warning)	Regional	Willing	Able
Scenario 5: Earthquake magnitude 7.0+ hits a Readiness Level II Partner Nation, results in a cholera outbreak	Permissive	II	Unpredictable	Regional	Willing	Able (limited)
Scenario 6: Readiness Level III Partner Nation experiences a magnitude 7.0+ earthquake and tsunami	Non-permissive	III	Unpredictable	Localized	Willing	Unable
Scenario 7: Multiple Readiness Level III Partner Nations in the midst of a regional conflict experience a deadly epidemic outbreak	Non-permissive	III	Unpredictable	Regional	Unwilling	Unable

These scenarios build from simplest to most complicated. In this table, the trade space consists of the scenario characteristics (e.g. Permissiveness, Class, etc.). We then modeled each scenario using a functional modeling tool. In the case below (see *Figure 2*), Innoslate® was used.

Figure 2 represents the simplest scenario (Scenario 1 from *Table 1*) of a hurricane approaching and making landfall. In an Action Diagram, the Actions are represented as rectangles. Rectangles with diamonds represent decision points and are special cases of the Action. Input/Output entities are represented by parallelograms where green indicates a trigger. Finally, resources are represented as purple hexagrams. By running the embedded discrete event simulator in Innoslate, we get the timeline shown in *Figure 3*.

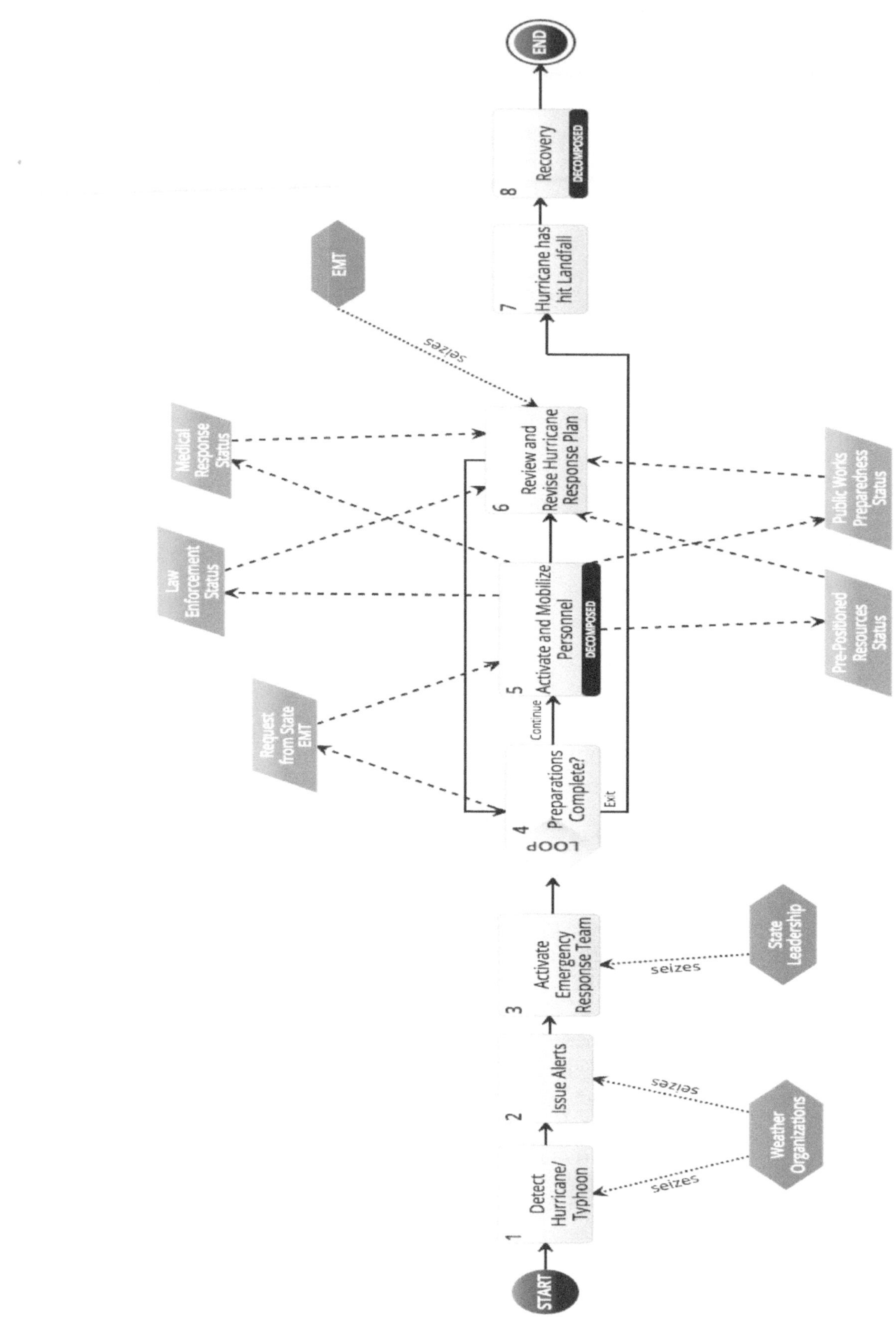

Figure 2. Action Diagram Model for Scenario 1 (Expected Hurricane)

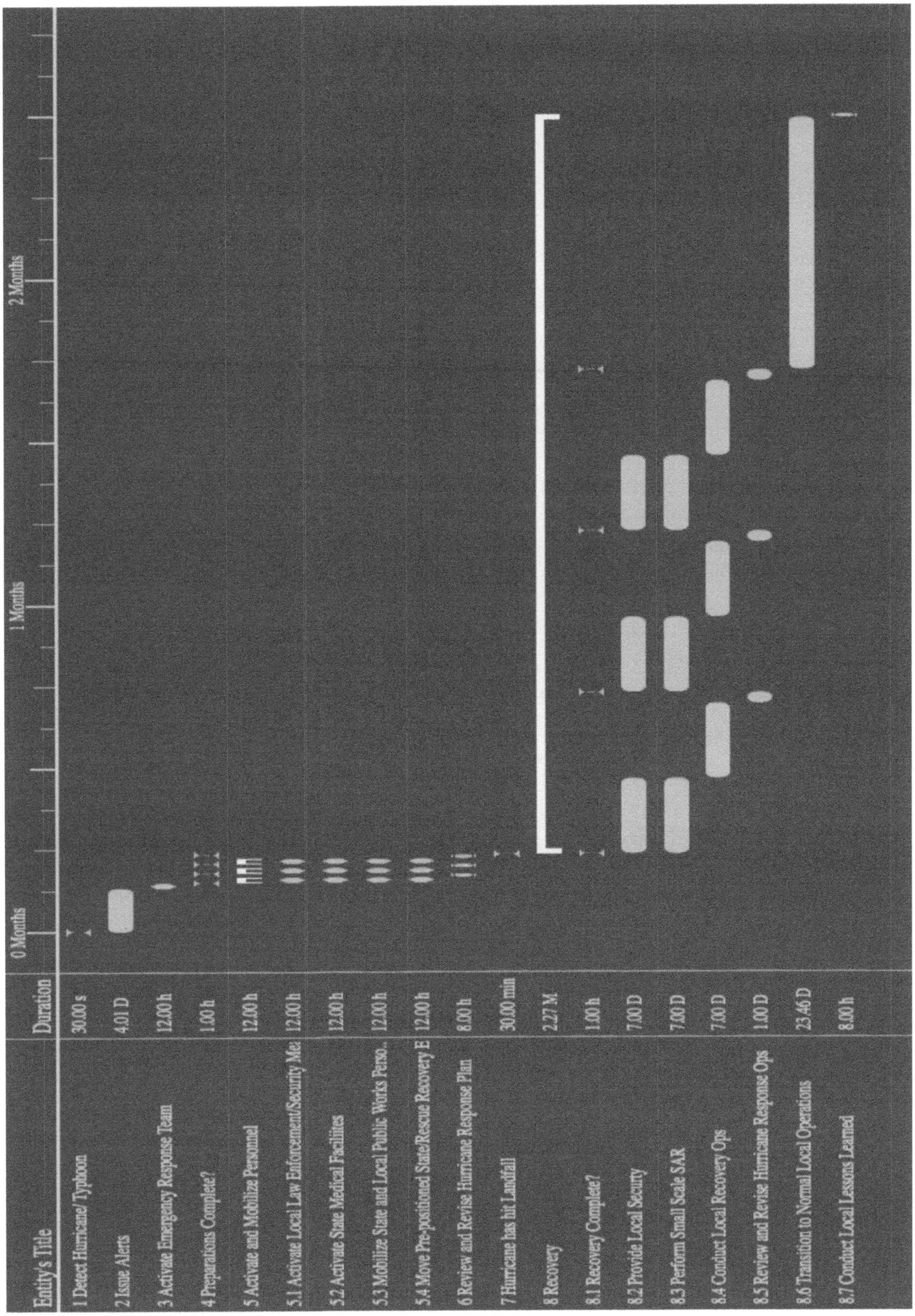

Figure 3. Timeline of Scenario 1

This timeline enables us to see how each step of the process is executed and provides the overall time for the process. It can also determine if logic errors have occurred or if resource limitations cause the simulation to fail. From this analysis, we can determine the performance constraints on the plan. We can add "what if" failure modes and effects analyses to this as well, thus making the overall process more robust.

Since many of these steps have variability, we can also add that variability to each step as a distribution, and then use Innoslate's Monte Carlo simulator to obtain the range of times and resources needed to successfully execute these processes.

We repeat this approach for each scenario with the following steps: 1) build the scenario model; 2) run the simulator and fix any problems with the execution of the model; 3) interpret the results to understand impacts on resources; 4) identify risks and decisions required; 5) develop risk mitigation approaches; and 6) document decisions and the rationale for the decisions.

As we model successive scenarios in *Table 1*, we see they build on one another, thus making the modeling easier and more rapid. We can then execute these models using discrete event and Monte Carlo simulations to obtain typical timelines, the range of times, costs and resource usage. These parameters become the basis of the key measures of effectiveness (MOEs).

When this functional analysis was completed, we only had to trace back to the original frameworks to identify any missing functional requirements. The result was a more comprehensive framework that forms the basis of the questions used in an on-line survey tool: ADRT.

Synthesis: Assessment of Disaster Readiness Tool (ADRT)

The purpose of the functional analysis is not just to verify that the general processes and procedures will work in the event of a crisis, but to also identify the common functions required by the supporting systems. From this analysis, a set of questions was developed for planners to determine their organization's specific needs.

ADRT is a survey tool, accessible online, that facilitates objective and measurable assessments used to help determine the capability, capacity, competency and self-sufficiency of a partner nation. ADRT also provides early warning information with links to the World Meteorological Organization (WMO), the National Oceanic and Atmospheric Administration (NOAA), the National Weather Service (NWS), and other monitoring agencies, as well as links to non-governmental organizations (NGOs), international non-governmental organization (INGOs), private voluntary organizations (PVOs), and other international emergency relief organizations for collaboration and coordination of efforts.

In short, ADRT provides:

- Disaster preparedness
- Management coordination
- Tracking of participating organizations to provide better coordination between organizations and their effort

This tool is available for free at: www.specinnovations.com/solutions/assessment-of-disaster-readiness-tool/.

An example of the questions in ADRT is shown in *Figure 4*. The question can be easily answered by knowledgeable disaster management personnel. Help information is provided as to what this means (as many users are expected to have English as a second language). Note the linkage to the source and associated framework. These references show the traceability back to the originating policies, which add credibility to the survey.

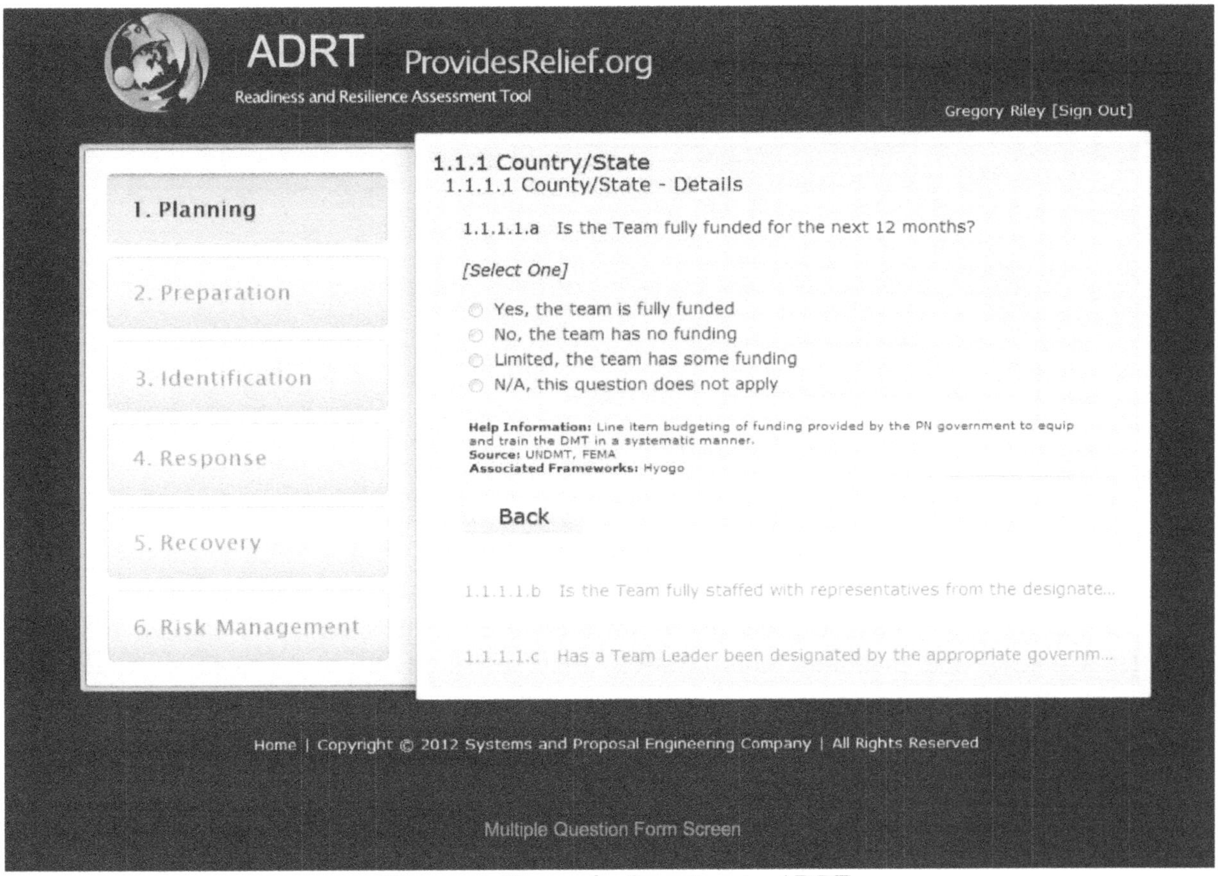

Figure 4. Example Question in ADRT

Figure 5 shows the results of using ADRT. The top left pie chart shows the percent of the questions that indicate the level of readiness. The top right chart shows the total score over time. As more assessments are made, users can see their progress. The bottom panel shows specific questions and their results, beginning with the needs. The information can be used to identify and justify the procurement of additional capabilities.

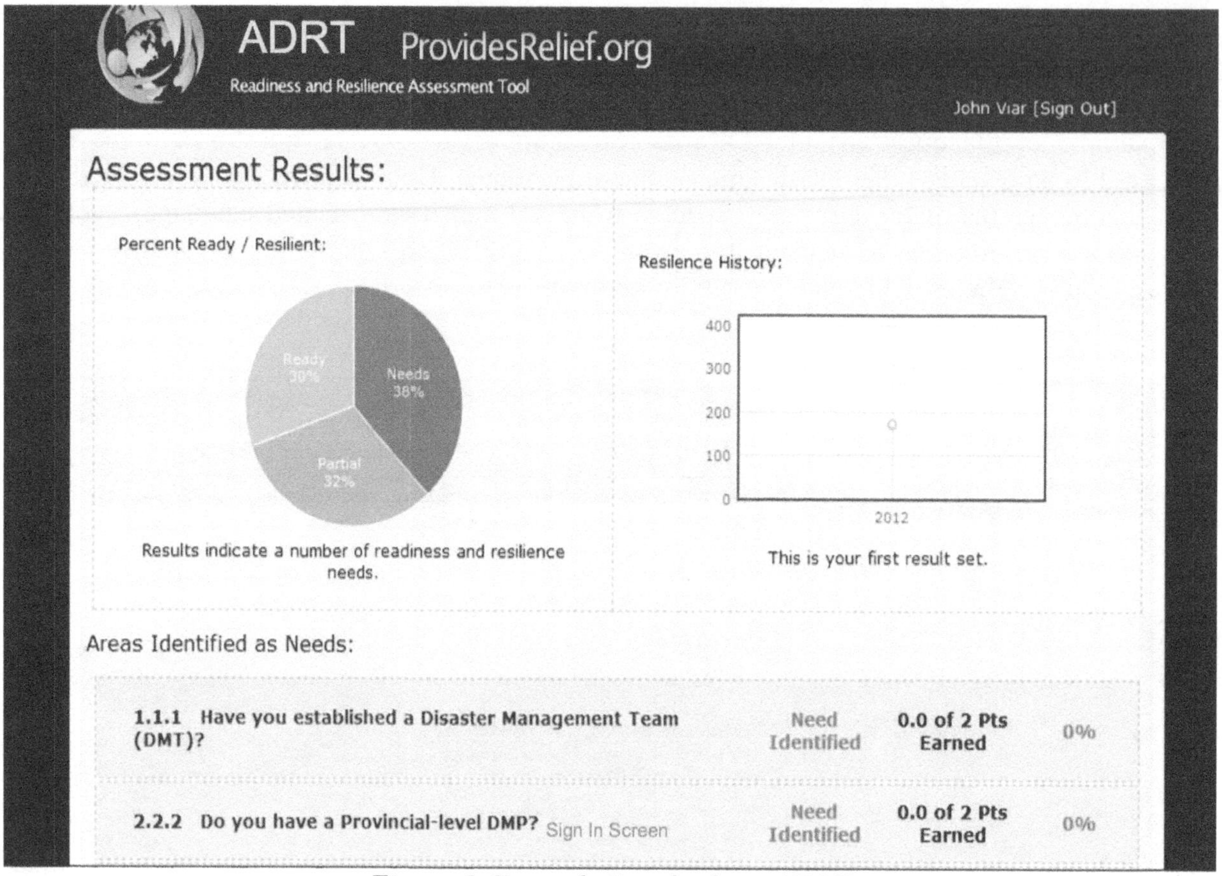

Figure 5. Example Results from ADRT

Conclusion

Since we can now store data securely on commercial and private clouds, we can use modeling tools (Innoslate®) and planning tools (ADRT) that utilize MBSE techniques anywhere and anytime to support disaster recovery. With access to worldwide databases, NGO's, PVO's and other international organizations can respond to threats or hazards anywhere in the world.

References

[i] DoD Architecture Framework 2.0 – A Guide to Applying Systems Engineering to Develop Integrated, Executable Architecture, Steven H. Dam, SPEC Innovations, 2015, p. 88

[ii] Yokohama Conference reference

[iii] "Hyogo Framework for Action 2005-2015: Building the Resilience of Nations and Communities to Disasters," World Conference on Disaster Reduction, 18-22 January 2005

[iv] Sendai conference reference

[v] Fixing Failed States – A Framework for Rebuilding a Fractured World, Ashraf Ghani and Clare Lockhart, Oxford University Press, 2008

[vi] "Interagency Conflict Assessment Framework (ICAF)," US Government Interagency Working Group, US Department of State's Office of the Coordinator for Stabilization and Reconstruction and USAID's Office of Conflict Management and Mitigation co-chairs, May 2008

[vii] "Reference Guide for OCHA's Strategic Framework 2010-2013," United Nations Office for the Coordination of Humanitarian Affairs (OCHA), 14 April 2010

[viii] "Measuring Progress in Conflict Environments (MPICE) - A Metrics Framework," United States Institute of Peace, 4 June 2010

Critical Infrastructure Protection and Recovery
A Common Recovery Model for Grid Systems in Blackout Scenarios

Casey Shull, Doctoral Student, Purdue University,
Team Leader Engineering / Operations Manager Crisis and Storm, Indianapolis Power & Light
Dr. Eric Dietz, Professor Purdue University / Director, Purdue Homeland Security Institute
Dr. Robert Kenley, Associate Professor Purdue University of Engineering
Practice of Industrial Engineering

Abstract. 25 years' experience observing and developing disaster and crisis recovery plans within the electric utility industry provides the impetus and knowledge to develop a software based single system strategic recovery algorithm for electric utilities utilizing system engineering techniques. The threat of system failure resulting from solar flares (Security Management), EMP (Savage, Gilbert, and Radasky) or ICS Cyber interruption (Weiss) on the electrical utility system can cascade up from the distribution system and or cascade down from the high voltage transmission system (U.S. Department of Energy). Systemic electrical system failure, regardless of the type of catastrophic event, cause the same socio-economic disasters within the region of electrical failure. This paper will address the concept of single system recovery techniques utilizing existing recovery methodologies, titled the Common Recovery Model (CRM), as an approach to hedge against and or recover from the potential of cascading failure of the electrical transmission and distribution system. The premise for the need of CRM is that one or all three of the threats are imminent, thus the necessity of electric utilities to develop CRM prior to the actual outage event.

Electric Utility System Reliability and Risk Assessment

Existing Regulations. Regulations in 2015 require electric utilities to report the quality of outage restoration through collection of yearly reliability indices. The data collected from each electrical outage during a calendar year reveals the speed at which an electric utility restores electricity to the rate payers by totaling the total minutes of outage per customer throughout the year. The reliability indices provide utilities with a measure to compare the reliability of each utility to one another within the same geographic region. The reliability indices encourage the use of automated devices to reduce the number of customers out of service faster than restoration through manual operations. The speed of restoration using automated devices prevents some of the electrical interruption from being classified as sustained outage, 5 minutes or greater by IEEE standard 1366-2012 (IEEE Std. 1366-2012), which are aggregated into the total number of electrical outage for a given event. The sum of events in a given calendar year are utilized to define the metrics of reliability.

Autonomous intelligently controlled devices improve the speed at which an electric utility can restore outage to affected customers to reduce time of outage to customers and improve reliability matrices. However, the automated devices can also be the proverbial "Achilles heel" to the electric utility due to the impact created by the three threats. Regulators recognize the potential for wide spread outage and are adding standards to compel electric utility operators to develop defense against the threats (University of Lincoln).

FERC 2016 approved the Geomagnetic Disturbances (GMD) Reliability Standard TPL-007-1 requiring grid operators to assess the vulnerability of their transmission system to GMD and prove their system meets certain performance requirements (FERC). The commission said: "While we recognize that scientific and operational research regarding GMD is ongoing, we believe that the potential threat to the Bulk Electric System warrants commission action at this time, including efforts to conduct critical GMD research." (FERC 15).

Sandia National Labs published a report for the U.S. Department of Energy (DOE) defining how to make the energy sector and electric grid resilient (USDOE). Indeed, defining methods to make the grid more resilient to attacks causing blackouts is a necessary endeavor (EEI). Resilience models, electrical load shedding and switching algorithms have been developed to define methodologies to make the grid risk averse and or minimize the significance of the outage. In addition, the economics of recovering from blackouts are detailed and well defined by computational methods by the IEEE (Arab, Khodael, Khator, and Han) for post disaster grid recovery.

What methods or technologies should be employed as an approach to recover from system wide blackout?

All electric utilities utilize the same technologies for operating and distributing electricity to consumers. Electrical transmission lines are interconnected through electrical switching devices that are operated utilizing software based systems vulnerable to the three threats. The map below provides an overview of the U.S. bulk transmission system.

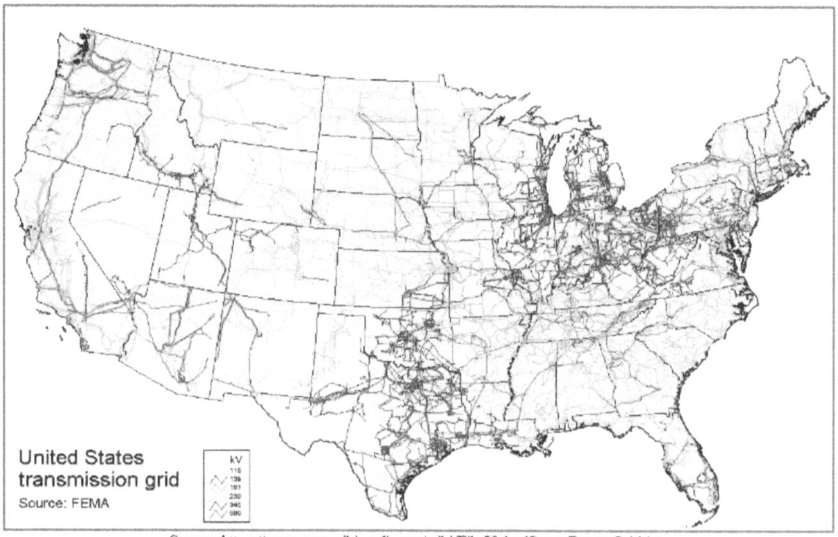

Source: https://commons.wikimedia.org/wiki/File:UnitedStatesPowerGrid.jpg

Fig. 1. Map of United States of America Electricity Grid - - National Energy Grids - Library Network Institute http://www.grid.org/globalenergy/library/national_energy_grid/united-states-of-america/americannationalelectricicitygrid.html

Recovery from regional or wide spread system failure require the same manual and technological approach to isolated outages. Regional or widespread blackout will require a specific approach and methodology for system wide recovery. Therefore, collecting strategies and data from existing electrical utility system recovery techniques will provide the information and opportunity to coalesce existing strategic approaches into the CRM methodology.

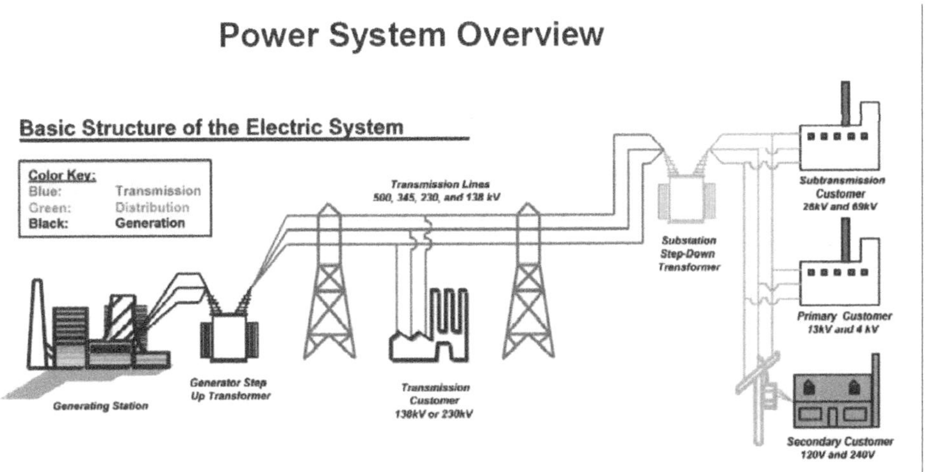

Figure 2-1. Power System Overview

Fig. 2. Report of the Commission to Assess the Threat to the United States from Electromagnetic Pulse (EMP) Attack

The system above provides an overview of a basic electrical system. The complexity of the overall system has been removed to identify the primary elements within the distribution system that would be affected by the three threats. Publications from the National Academy of Sciences (Carlowicz and Lopez) provide insight regarding the intensity of EMP and Solar Flares resulting in the impact and loss of functionality of the various automated control devices. Blackouts from one of the three threats would create the inability for utility operators to control the distribution system causing loss of service to other areas of critical infrastructure.

The President of the United States issued Executive Order – Coordinating Efforts to Prepare the Nation for Space Weather Events. The Executive Order Section 4 part (d) "The Secretary of Energy shall facilitate the protection and restoration of the reliability of the electric power grid during a presidentially declared grid security emergency associated with a geomagnetic disturbance pursuant to 16U.S.C 824o-1" (Ex.: 13744 issued October 13, 2016).

Common Recovery Model

The application of system engineering techniques to understand the event, develop strategies of threat detection and strategies for recovery, define the "critical path" or CRM utilizing existing or similar recovery techniques to reduce and or prevent catastrophic loss will delay or prevent socio-economic chaos. The following elements within an electric utilities system should be defined by priority of need for the public welfare utilizing MBSE.

Identify clusters of critical infrastructure (CI): Electrical Substations (1), Hospital, (2)

Critical Infrastructure Cluster			
Priority		Lat	Long
1	Substation		
2	Hospital		
3	Food Water		
4	Security		
5	Shelter		

Table 1 Basic table listing priorities to use MBSE to develop CRM

Food and Water (3), Emergency and Security (4), Shelter (5). The assignment of priorities can vary and or increase based upon emergency operators and responder's development of priorities.

The list of CI can increase and change based upon the clusters of CI in an area relative to the population. The following is a basic list to provide the concept and methodology using MBSE to define the Common Recovery Model for each utility. Utilizing the table above to develop CRM using MBSE techniques and Architecture will provide a model to recovery from regional or wide spread electrical outage.

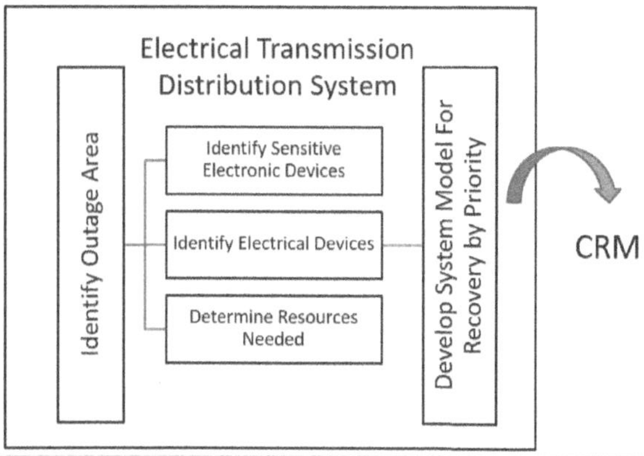

Figure 3 Basuc Concept to develop CRM

System engineering tools will develop CRM to identify the critical path to electrical recovery based upon the priority given to each CI that is in proximity to the electrical system that can provide recovery to the largest number of critically grouped priorities. All electric utilities require the same system engineering approach, hence the title CRM, to recovery regardless of the number of CI served and the number of electrical facilities in the system. Electric utilities can develop CRM utilizing tools, techniques and architecture within various MBSE software products.

Conclusion: Electric utility operators, industry experts, various government agencies and most recently President Obama have issued reports, publication and executive orders providing recommendations and orders to defend or make the "grid" more resilient. The threat is imminent, thus a method to recover from regional or wide spread electrical outage must be developed.

All electric utilities have installed and developed autonomous electrical systems to decrease the duration of electrical outage for consumers. The autonomous devices provide the fastest response to minor electrical outage, however, the devices are also more susceptible to one of the three threats. Therefore, all electric utilities can utilize a common method for recovery, the Common Recovery Model.

Works Cited

Arab, Ali, Amin Khodael, Suresh Khator, and Zhu Han. *Electric Power Grid Restoration Considering Disaster Economics*. New York: IEEE, 2016. *IEEE website*. Web. 1 Oct. 2016. <http://ieeexplore.ieee.org.ezproxy.lib.purdue.edu/document/7395278/?reload=true>.

Carlowicz, Michael J., and Ramon E. Lopez. *Storms From the Sun: The Emerging Science of Space Weather*. Washington D.c.: National Academy of Sciences, 2002. *National Academy of Sciences website*. Web. 10 May 2016. <www.nationalacademies.org>.

Edison Electrical Institute. *Before and After the Storm*. Washington D.c.: EEI, 2014. *EEI website*. Web. 10 Aug. 2016. <http://www.eei.org/issuesandpolicy/electricreliability/mutualassistance/Documents/BeforeandAftertheStorm.pdf>.

Exec. Order No. Executive Order number. EX.: 13744 issued October 13, 2016, 3 C.F.R. https://www.federalregister.gov/. (2016). Print.

Federal Energy Regulatory Commission. *Reliability Standard for Transmission System Planned Performance for Geomagnetic Disturbance Events*. 18 CFR Part 40 [Docket No. RM15-11-000]. Washington D.C.: FERC, 2015. *FERC*. Web. 1 Oct. 2016. <https://www.ferc.gov/whats-new/comm-meet/2015/051415/E-1.pdf>.

Harwood, Mark. *Sun cycle threatens electric grid*: Security Management, 2012. *Security Management website*. Web. 20 July 2016. <https://sm.asisonline.org/Pages/Sun-Cycle-Threatens-Electric-Grid.aspx>.

High-Voltage Transformer Substations . Washington D.C.: Congressional Research Service, 2014. *Federation of American Scientists website*. Web. 10 Oct. 2016. <http://fas.org/sgp/crs/homesec/R43604.pdf>.

IEEE Std. 1366-2012. *IEEE Guide for Electric Power Distribution Reliability Indices*. New York: IEEE, 2012. *IEEE website*. Web. 15 April 2016. <>.

Savage, Edward, James Gilbert, and William Radasky. *The Early-Time (E1) High-Altitude electromagnetic Pulse (HEMP) and Its Impact on the U.S. Power Grid*. Oak Ridge, Tennessee: Metatech, 2010. *Ferc website*. Web. 20 July 2016. <https://www.ferc.gov/industries/electric/indus-act/reliability/cybersecurity/ferc_meta-r-320.pdf>.

University of Lincoln. "New research warns world to prepare for power outages." Home page. *Science Daily webpage*. Science Daily, 27 Jan. 2014. Web. 1 Oct. 2016. <https://www.sciencedaily.com/releases/2014/01/140127093033.htm>.

U.S. Department of Energy. *Advanced Simulation for Analysis of Critical Infrastructure: Abstract Cascades, the Electric power grid, and Fedwire*. Springfield, VA: Sandia National Laboratories, 2004. *Sandia.gov website*. Web. 1 May 2016. <http://www.sandia.gov/nisac/wp/wp-content/uploads/downloads/2012/04/simulation-and-analysis-of-cascading-failure-in-critical-infrastructure.pdf>.

---. *Conceptual Framework for Developing Resilience Metrics for the Electricity, Oil, and Gas Sectors in the United States*. Arlington, VA: Sandia National Laboratories, 2014. *Energy.gov webpage*. Web. 10 Aug. 2016. <http://energy.gov/sites/prod/files/2015/02/f20/EnergyResilienceRpt-Sandia-Sep2014.pdf>.

Weiss, Joseph. "Implications of the 2015 Ukrainian Cyber Attack to Critical Infrastructures ." *INCOSE* (2016): 1+. Print.

Space Weather Policy and Effects

Seth Jonas, sjonas@ida.org
Eoin McCarron, emccarron@ida.org
William Murtagh, william.murtagh@noaa.gov

Abstract

Space weather events have the potential to negatively affect technology and energy infrastructures that are essential contributors to national security, economic vitality, and daily functioning. The term "space weather" refers to the naturally occurring, dynamic conditions of the space environment that arise from emissions from the sun, which include solar flares, solar energetic particles, and coronal mass ejections. These emissions can interact with Earth, potentially degrading, disrupting, or damaging electric power systems; satellite, aircraft, and spacecraft operations; telecommunications; position, navigation, and timing services; and other technologies and infrastructures. As these technologies become increasingly ubiquitous and interdependent, the United States—and, indeed, the world—faces greater risks from the threats posed by space weather events. Critical infrastructures make up a diverse, complex, interdependent system of systems in which the failure of one could cascade to another. Understanding this threat and the effects of space weather on critical systems is crucial to enhancing preparedness and building resilience. This article (1) provides readers with an overview of the space weather phenomena most likely to degrade, disrupt, or damage critical infrastructures; (2) describes the US National Oceanic and Atmospheric Administration's process for communicating space weather conditions and their possible effects on critical systems; and (3) detail current Federal and international efforts to enhance space weather preparedness.

Introduction

Space weather refers to the dynamic conditions of the space environment that arise from emissions from the sun and can affect the reliability and performance of space- and ground-based technological systems, as well as endanger human health (National Science and Technology Council 2015a). Space weather results from a range of solar phenomena, including coronal mass ejections (CMEs), solar flares, solar energetic particle events, and solar winds. CMEs are explosions of plasma from the sun that can cause geomagnetic storms and enhanced aurora when they interact with Earth's magnetic field. Solar flares, commonly associated with sunspots, are brief emissions of high-energy electromagnetic radiation. Solar energetic particle events refer to atoms ejected from the sun that are typically associated with solar eruptions, while solar wind is the stream of plasma that the sun continually emits. Together these phenomena are responsible for the primary types of space weather events: radio blackouts, solar radiation storms, and geomagnetic storms (also referred to as geomagnetic disturbances or GMDs). These events can degrade, disrupt, or damage critical infrastructure, including the electric power grid; position, navigation, and timing (PNT) services such as the global positioning system (GPS), global navigation satellite system (GNSS), and long-range navigation (LORAN); communications; satellites; aviation; and space flight.

As technological systems and infrastructures become increasingly interdependent and ubiquitous, space weather events pose a growing risk to the world. Space weather events could cause continental-scale cascading failures across interconnected infrastructure systems, such as the bulk-power system, and could cause errors in PNT service which are relied on by users worldwide. Understanding the threat of space weather events, their potential effects, and the vulnerability of infrastructure and technology systems is critical to improving space weather preparedness and building resilient systems.

Space Weather Effects

Space weather can present a significant risk to critical infrastructure operations and human health. Changes in the dynamics and composition of the solar wind interacting with Earth's magnetic field lead to geomagnetic storms. An effect of significant concern is electrical currents induced in long conductors such as the electric power grid, gas and oil pipelines, and railways. These geomagnetically induced currents can disrupt service or damage components of the electrical power grid, corrode pipelines, and disrupt rail service.

Such effects have been observed in the past across the globe. The 1859 Carrington event is often considered to be the largest geomagnetic storm on record (Tsurutani et al. 2003; Pulkkinen et al. 2012). Although the storm occurred at a time before widespread electrification of America, the Carrington event disrupted the primary electronic system of the day, which was the land-based telegraph network. Telegraph operators on the East Coast of the United States reported outages as equipment sparked and in some places caused combustion (Boteler 2006). Observers in Hawaii and Cuba reported auroras, a visible manifestation of space weather that generally occurs only at much higher latitudes (Tsurutani et al. 2003). If an event of this magnitude occurred today, it could cause widespread and long-term power outages. According to one estimate, Carrington-scale event could cause $1 trillion to $2 trillion in damages over a year (National Research Council 2008), According to another estimate, damages in North America alone could be between $0.6 and $2.6 trillion (Lloyd's and Atmospheric and Environmental Research, Inc. 2013).

The Carrington event is an important historical reference point, but recent examples may be more useful when describing the potential effects of space weather on critical infrastructure. In March 1989, a geomagnetic storm caused the collapse of the Canadian Hydro-Quebec electric power grid, leaving millions of customers without power and demonstrating the vulnerability of the electric power grid to space weather (National Research Council 2008). The storm also damaged grid components in the United States (North American Electric Reliability Corporation1990; Oak Ridge National Laboratory 1991). Following the 1989 event, Hydro-Quebec invested over CAD $1.2 billion in monitoring and mitigation technology (Government of Canada, Office of Critical Infrastructure Protection and Emergency Preparedness 2002). In the United States, the US Nuclear Regulatory Commission issued information regarding the storm's effects and North

American Electric Reliability Corporation called on the US National Weather Service to improve its forecasting of geomagnetic storms (Nuclear Regulatory Commission 1990; North American Electric Reliability Corporation 1990). Over a decade later, the less-intense Halloween Storms of 2003 had significant effects on modern technological systems and put human health at risk. For example, the events prompted NASA to direct astronauts on the International Space Station to seek shelter from the increased radiation levels, airlines to reroute flights over polar areas to lower latitudes due to communication problems and a potential increase in radiation-related health risks. Further, geomagnetically induced currents caused electrical transformer problems in Northern Europe and an electric power grid failure in Sweden (National Oceanic and Atmospheric Administration 2004).

Other types of space weather can also affect infrastructure systems. For example, in November 2015, radio emissions associated with a solar flare affected the operation of Swedish radar stations used for air traffic control. This event resulted in flight delays and disruption in Swedish airspace for nearly an hour (Luftfartsverket 2015). In January 2014, NASA reported that a rocket launch to resupply the international space station was scrubbed due to "an unusually high level" of solar energetic particles, which could potentially affect the rocket's avionics (NASA 2014).

Communicating the Risk of Space Weather

To help communicate the risk of space weather, the U.S. National Oceanic and Atmospheric Administration (NOAA) Space Weather Prediction Center (SWPC) issues space weather forecasts for the United States, which are available via free subscription on the SWPC website (National Weather Service (NWS) 2016). The forecasted watches and warnings are communicated using the NOAA Space Weather Scales, which indicate event severity, frequency, and possible effects on humans and technology systems (NWS 2011). The NOAA Space Weather Scales rank geomagnetic storms (G-scale), solar radiation storms (S-scale), and radio blackouts (R-scale) on a numbered scale based on their severity. Versions of these scales have been adopted across the globe, and refinements have been made for effects based on geomagnetic latitude and differences in technology systems. The scales are similar to others that convey the severity of natural disasters (e.g., the Saffir-Simpson hurricane wind scale and the Richter magnitude scale).

The NOAA Space Weather Scales categorize geomagnetic storms from G1 (minor) to G5 (extreme). According to the scales, on average, there are around 2,600 geomagnetic storms every solar cycle, where a solar cycle lasts 11 years on average. The most intense storms, G4 (severe) and above, occur much less frequently: approximately 100 per cycle. Storms described by the scales as severe or extreme can cause voltage control problems in power systems, the collapse of the electric power grid, and even transformer damage. Geomagnetic storms can also affect space-based systems, causing spacecraft surface charging and disrupting spacecraft tracking and orbits. Other effects include currents in pipelines, the degradation of PNT services, and disruptions to high-frequency radio communications.

Similar to geomagnetic storms, the NOAA Space Weather Scales rank solar radiation storms from S1 (minor) to S5 (extreme). There are on average 88 solar radiation storms per solar cycle. Events characterized as S4 (severe) occur approximately three times per cycle and S5 (extreme) occur even less frequently. Radiation storms described by the scales as severe or extreme can have a range of negative impacts on human health, satellites, and communication systems. Events of this scale pose a radiation hazard for astronauts preforming tasks outside their spacecraft and could expose passengers and crew in high-flying aircraft to increased radiation. Solar radiation storms of these magnitudes can also disrupt or even disable satellites, causing problems with data storage and memory devices, noise on imaging systems, and degrading solar panel efficiency. They can also cause the temporary blackout of high-frequency radio communications across polar regions and increase the potential for navigation errors.

The NOAA Space Weather Scales categorize radio blackouts from R1 (minor) to R5 (extreme). These events occur on average more than 2,500 times per solar cycle, with minor events accounting for the vast majority of these instances. Events rated R4 (severe) or higher occur approximately eight times per cycle and can severely degrade high-frequency radio communications and PNT services.

While the NOAA Space Weather Scales are an effective tool for communicating the severity of a space weather event and its potential effects, they are limited in their ability to describe extreme space weather. The "extreme" categories of the scales have no upper boundary. An extreme event could range in intensity from the March 1989 event that caused an electrical blackout in Quebec, Canada, to the significantly more intense Carrington event of 1859 (Tsurutani et al. 2003; Siscoe et al. 2006; Pulkkinen et al. 2012).

A scale that links geomagnetic storm intensity to specific effects on infrastructure systems does not currently exist. The US National Centers for Environmental Information does, however, host a database of geomagnetic storm intensity dating back to 1933 (National Oceanic and Atmospheric Administration 2016). This database refers to the Kp index, a global index derived from a weighted average of K-indices from a network of worldwide geomagnetic observatories. The K-indices are 3-hour-long measurements of the maximum fluctuations of the horizontal components of the Earth's magnetic field observed on a magnetometer at a specific location. The Kp index can, in turn, be mapped back to the NOAA Space Weather Scales. A G5 event is classified as having a Kp value of 9 (the index maximum); a G4, a value of 8 or 9; a G3, a value of 7; a G2, a value of 6; and a G1, a value of 5.

Enhancing Space Weather Preparedness

The NASA Authorization Act of 2010 recognizes that "space weather events pose a significant threat to modern technological systems" and that "the effects of severe space weather events on the electric power grid...would have significant societal, economic, national security, and health impacts" (42 United States Code Section 18388(a)). The Act also orders the Director of the White House Office of Science and Technology Policy to "improve the Nation's ability to prepare, avoid, mitigate, respond to, and recover from the potentially devastating impacts of space weather events" (42 United States Code Section 18388(b)(1)).

In 2013, the US Federal Energy Regulatory Commission issued Order 770 that directed the North American Electric Reliability Corporation to develop a two-stage reliability standard to address the effects of geomagnetic storms on the reliable operation of the bulk-power system (Federal Energy Regulatory Commission 2013). In June 2014, the commission adopted the first stage, which orders power grid owners and operators to implement operating plans and procedures that mitigate the effects of geomagnetic storms (18 Code of Federal Regulations Part 40). The second phase, which is still under review, will require power grid owners and operators to assess the impact of identified benchmark geomagnetic storms on the bulk-power system.

In 2014, the White House, building on previous interagency efforts to develop a coordinated federal government response to space weather, established the Space Weather Operations, Research, and Mitigation (SWORM) Task Force, co-chaired by the White House Office of Science and Technology Policy, the Department of Homeland Security, and the Department of Commerce (Jonas et al. 2016). In 2015, the task force developed the "National Space Weather Strategy" and the accompanying "National Space Weather Action Plan" in, which "articulate how the federal government will work to enhance national space-weather preparedness by coordinating, integrating, and expanding existing policy efforts; engaging a broad range of sectors; and collaborating with international counterparts" (White House Office of Science and Technology Policy 2015). The strategy identifies six high-level goals: (1) establish benchmarks for space weather events; (2) enhance response and recovery capabilities; (3) improve protection and mitigation efforts; (4) improve assessment, modeling, and prediction of impacts on critical infrastructure; (5) improve space weather services through advancing understanding of forecasting; and (6) increase international cooperation (National Science and Technology Council 2015a). The Action Plan identifies approximately 100 distinct activities and related timelines and deliverables that will help achieve these goals. Since the October 2015 release of the Strategy and Action Plan, U.S. agencies have worked cooperatively to implement the goals and activities outlined in these documents.

Upon completion, goal 1 of the strategy will identify a set of benchmarks that define characteristics and conditions of extreme space weather events for use by systems engineers and planners to help inform system development. The benchmarks will provide a point of reference from which to inform the development of mitigation procedures, enhance response and recovery planning, and understand risk (National Science and Technology Council 2015b). The interagency teams developing the benchmarks are basing their work on current scientific understanding and historical observations with the aim of providing clear and consistent descriptions of the physical parameters of space weather phenomena. The benchmarks will address different types of space weather events and will identify a range of event magnitudes and recurrence rates, including 1-in-100-year events and the theoretical maximum. These benchmarks will focus on the physical phenomena of extreme space weather, not a particular effect on systems. When completed, these benchmarks could serve as input to vulnerability assessments, threshold points for action, and tools to assist systems engineers.

The enacted 2015 Fixing America's Surface Transportation Act (Public Law 114-94) identifies geomagnetic storms as a potential cause of a power grid security emergency and amends the Federal Power Act (16 United States Code Section 824 et seq.), giving the Secretary of Energy, upon the declaration of an emergency by the President, authority to issue orders to protect the reliability of critical electric infrastructure (16 United States Code Section 824(O-1)(b)(1)). Recent legislation introduced in the US Senate Committee on Commerce, Science, and Transportation—the Space Weather Research and Forecasting Act

(S.2817)—also hopes to address the threat of space weather and calls for a concerted federal effort to enhance preparedness and resilience.

There are also significant international efforts to address the threat of space weather. For example, in 2012, the UK government, through the Ministry of Defense, issued a formal response to the UK Parliament Defense Committee acknowledging the need to address the threat posed by space weather and man-made electromagnetic disturbances (e.g., an electromagnetic pulse) (UK Ministry of Defense 2012). In 2012, the UK also added space weather to its National Risk Register of Civil Emergencies, an unclassified version of the National Risk Assessment, an assessment of the risks of civil emergencies facing the UK over the next 5 years (UK Cabinet Office 2012). The latest edition of the National Risk Register for Civil Emergencies (2015) identifies space weather as a risk and describes a reasonable worst-case scenario (UK Cabinet Office 2015). In 2014, the UK established the Met Office Space Weather Operations Centre, a 24/7 operations center that provides space weather forecast and warnings of the "impacts of space weather on UK services and infrastructure into Government and responder communities" (UK Met Office 2016).

The US space weather strategy and action plan can further international awareness and increase efforts; together, they call for a collaborative and international approach for preparing for extreme space weather events. In response to the publication of the action plan, the US Department of State and the Secure World Foundation convened a panel discussion on space weather as a global challenge where US government officials joined experts from the UK, the United Nations (UN) World Meteorological Organization, the UN Committee on the Peaceful Uses of Outer Space, and the North Atlantic Treaty Organization (NATO) to discuss ways to further international collaboration on space-weather mitigation efforts, research, forecasting, and risk management activities. (Department of State 2016; Secure World Foundation 2016).

Conclusion

A large geomagnetic storm could cause cascading failures that disable portions of the electric power grid for hours to days (or even longer) and cause harm to the economy and human health. Outages of other important infrastructures, including commercial air travel, satellites (space-based PNT and telecommunications), and mobile communications and pipelines, could lead to cascading failures across infrastructure sectors. Recent efforts in the United States and abroad, including the White House "National Space Weather Strategy" and "National Space Weather Action Plan," have sought to address space weather's negative impacts by coordinating research and mitigation efforts. Moving forward, systems engineers should engage with and monitor the activities, specifically the development of strategy and action plan benchmarks and the NOAA Space Weather Scales. Building awareness of the negative effects of space weather within the systems engineering community will help encourage the design and development of robust and resilient infrastructure around the world.

References

Boteler, D. H., 2006. "The Super Storms of August/September 1859 and Their Effects on the Telegraph System." *Advances in Space Research* 38, 159–172.

Department of State. 2016. "Space Weather as a Global Challenge." http://www.state.gov/e/oes/sat/spaceweather/.

Federal Energy Regulatory Commission. 2013. "Reliability Standards for Geomagnetic Disturbances." Order No. 779. http://www.ferc.gov/whats-new/comm-meet/2013/051613/E-5.pdf.

Jonas, S., and E. D. McCarron, 2016, "White House Releases National Space Weather Strategy and Action Plan." *Space Weather* 14, 54–55. doi:10.1002/2015SW001357

Luftfartsverket. 2015. "Full Capacity after 90 Minutes Radar Loss." https://www.lfv.se/en/news/news-2016/full-capacity-after-90-minutes-radar-loss.

Lloyd's and Atmospheric and Environmental Research, Inc. 2013. "Solar Storm Risk to the North American Electric Grid." https://www.lloyds.com/~/media/lloyds/reports/emerging%20risk%20reports/solar%20storm%20risk%20to%20the%20north%20american%20electric%20grid.pdf.

NASA. 2014. "Cygnus Launch Now Set for Thursday; Light Duty Day on Station." http://www.nasa.gov/content/cygnus-launch-now-set-for-thursday-light-duty-day-on-station.

National Research Council. 2008. "Severe Space Weather Events—Understanding Societal and Economic Impacts: A Workshop Report." Washington, DC: National Academies Press.

National Science and Technology Council. 2015a. "National Space Weather Strategy." https://www.whitehouse.gov/sites/default/files/microsites/ostp/final_nationalspaceweatherstrategy_20151028.pdf.

———. 2015b. "National Space Weather Action Plan." https://www.whitehouse.gov/sites/default/files/microsites/ostp/final_nationalspaceweatheractionplan_20151028.pdf.

National Oceanic and Atmospheric Administration. 2004. "Service Assessment: Intense Space Weather Storms October 19–November 07, 2003." http://www.nws.noaa.gov/os/assessments/pdfs/SWstorms_assessment.pdf.

———. 2016. "Geomagentic Kp and Ap Indices." http://www.ngdc.noaa.gov/stp/GEOMAG/kp_ap.html.

North American Electric Reliability Corporation (NERC). 1990. "NERC Position Statement on Solar Magnetic Disturbance Forecasting." Approved by the Board of Trustees July 9, 1990. http://www.nerc.com/files/1989-Quebec-Disturbance.pdf.

Nuclear Regulatory Commission. 1990. "Information Notice No. 90-42: Failure of Electrical Power Equipment Due To Solar Magnetic Disturbances." Washington, DC: Office of Nuclear Reactor Regulation. http://www.nrc.gov/reading-rm/doc-collections/gen-comm/info-notices/1990/in90042.html.

Oak Ridge National Laboratory. 1991. "Electric Utility Experience Industry with Geomagnetic Disturbances." ORNL-6665. http://web.ornl.gov/~webworks/cpr/v823/rpt/51089.pdf.

Pulkkinen, A., E. Bernabeu, J. Eichner, C. Beggan, and A. W. P. Thomson. 2012. "Generation of 100-Year Geomagnetically Induced Current Scenarios." *Space Weather* 10, S04003. doi:10.1029/2011SW000750.

Secure World Foundation 2016. "Space Weather as a Global Challenge." http://swfound.org/events/2016/space-weather-as-a-global-challenge.

Siscoe, G., N., U. Crooker, and C. R. Clauer. 2006. "DST of the Carrington Storm of 1859." *Advances in Space Research* 38: 173–179. doi:10.1016/j.asr.2005.02.102.

National Weather Service. 2011. "NOAA Space Weather Scales." http://www.swpc.noaa.gov/sites/default/files/images/NOAAscales.pdf.

*———. 2016. "Space Weather Enthusiasts Dashboard." http://www.swpc.noaa.gov/communities/space-weather-enthusiasts.

*______. 2016. "Subscription Services." http://www.swpc.noaa.gov/content/subscription-services.

Tsurutani, B. T., W. D. Gonzalez, G. S. Lakhina, and S. Alex. 2003. "The Extreme Magnetic Storm of 1–2 September 1859." *Journal of Geophysical Research* 108(A7): 1268. doi:10.1029/2002JA009504.

UK Cabinet Office. 2012. "National Risk Register of Civil Emergencies." https://www.gov.uk/government/uploads/system/uploads/attachment_data/file/211858/CO_NationalRiskRegister_2012_acc.pdf.

UK Cabinet Office. 2015. "National Risk Register of Civil Emergencies." https://www.gov.uk/government/uploads/system/uploads/attachment_data/file/419549/20150331_2015-NRR-WA_Final.pdf.

UK Met Office. 2016. "The Met Office and Space Weather." http://www.metoffice.gov.uk/publicsector/emergencies/space-weather/met-office-role.

UK Ministry of Defense. 2012. "Developing Threats: Electro-Magnetic Pulses (EMP): Government Response to the Committee's Tenth Report of Session 2010–12." http://www.publications.parliament.uk/pa/cm201012/cmselect/cmdfence/1925/192504.htm.

White House Office of Science and Technology Policy. 2015. "Enhancing National Preparedness to Space-Weather Events." https://www.whitehouse.gov/blog/2015/10/28/enhancing-national-preparedness-space-weather-events.

Cybersecurity and the Critical Vulnerability of the Dutch Water Control Systems

Berber de Liefde, berber.de.liefde@gmail.com
Jan de Liefde, jandeliefde@gmail.com

Abstract
Everybody can imagine what will happen, when the Internet is attacked by a cyber-attack or when it is not possible anymore to get money out the ATM. The consequences of a cyber-attack for water barriers, locks and sluices, retaining structures, pumping stations and tunnels are not that visible for most Dutch people. Depending of the season and the level of the water the mainland will be flooded or will be dried if it's not protected by retaining structures or pumping stations. Transportations on water or road will stop by locks, sluices and tunnels. The effect will be that stores and companies aren't provided with stocks anymore. In the worst case the import and export will be on hold, this will have a major impact at the Dutch economy, because they depend on the trading.

This article will start with the cyber treatments in relation to many critical water and transportation systems of the Netherlands. The next part will give an idea which measures the Dutch government prescribed to the different contracting parties who are involved during design, realisation, operation and maintenance of the infrastructure systems. The last part will give you a glimpse in the consequences of these measures for each of the phases development, realisation and maintenance executed by private companies.

Introduction
Critical infrastructure as we defined it in this article is the Infrastructure that is part of the four infrastructure networks within the Netherlands: water management, main waterways, highways and railways. What make these networks so critical for the Netherlands? They have a large geographical scale and they are connected with large information networks across the Netherlands. They are one of the most important enablers of the normal life in the Netherlands.

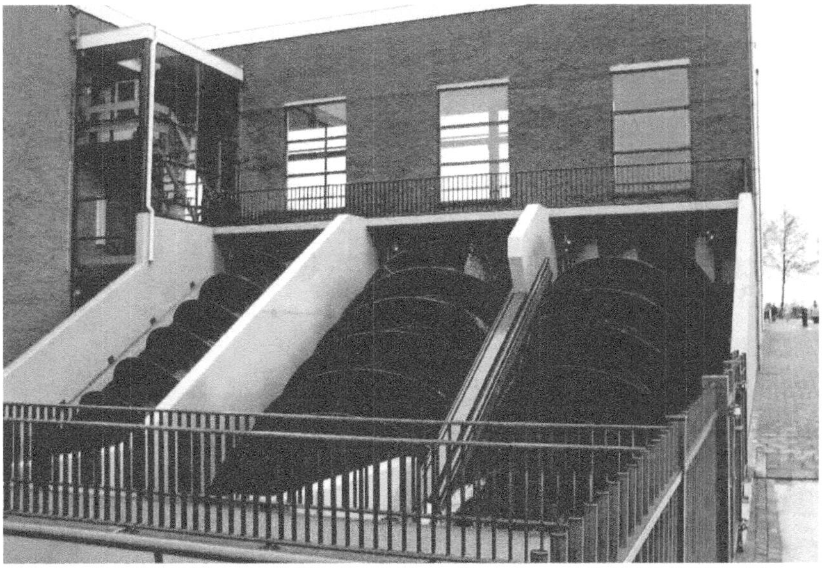

Figure 1 Kinderdijk – Pumping station

Figure 2 – Maeslandtkering, storm surge barrier

Water management is a complex network of dikes, moveable water barriers, locks, pumping stations and they are mostly connected to central monitoring and controlling centers.
The water management network main goals are:

- Protect us against the water of the North Sea and the water from the large rivers like Rijn and Maas.
- Guarantee delivering of fresh water to water distribution companies and agriculture

Regulations regarding to our high Water defense system is stated as a law in the Waterwet [5].
For example, the protection level (expressed in a failure rate) against high water levels for different area's is a law statement [6].

Waterways are a Complex System of Systems build up on so called corridors which are monitored and controlled from so called Nautical control centres, Corridors are a combination of channels and rivers where by different water levels they are separated by sluices. Crossings with land traffic are realized by moveable and non-moveable bridges, tunnels and aquaducts. The moveable, non-moveable bridges are part of waterway network while tunnels are part of our highway network. Bridges, sluices, radarposts and dynamic water traffic signs are mostly controlled and monitored from Nautical control centers at land- and regional level.
The waterways network main goals are

- Provide high speed corridors for commercial water traffic regarding to the availability of open bridges and sluices on agreements between commercial parties and the government.
- Provide interconnections between water recreation areas for recreation boats.

Highways are network of traffic ways operated and maintained by the government. They are monitored and controlled from five traffic control centres. The traffic centres control and monitor the dynamic traffic management systems and tunnels.
The highway main goal:

- Provide a safe and reliable connection between cities;

Railway networks are a complex network of rail connections between cities and areas for the transportation of people and goods by railway stations, yards, and maintenance yards. There is separation by law of local (metro and tram) and state railways (high speed and regional).
They are mostly operated and maintained by one party and used by many different parties. Information about travel information is shared by all parties.
The different railway networks are monitored and controlled from rail traffic centers across the country mostly posted. Information about traffic information is shared by different operating companies.
There are European safety regulations defined for these railway systems and their components. networks. From government side (local and state) there are availability agreements stated.
The railway main goals are:

- Provide a scheduled and safe connection between different interconnection points;
- Provide a comfortable connection between different interconnection points;
- Provide reliable travel information.

As mentioned earlier these Infrastructure systems heavily depends on interconnected information networks. Not only within one organization but especially between different organizations and among people. Examples are: operating companies, maintenance companies, ship crew, travelers, carrier organizations, etc.

So therefore information should be reliable. Interconnected networks makes this information vulnerable to cyber-attacks.

In Cybersecurity 2015 [1] it is stated:

> ***Cyber security*** *refers to efforts to prevent damage caused by disruptions to, breakdowns in or misuse of ICT and to repair damage if and when it has occurred.*
>
> *Such damage may consist of any or all of the following: reduced reliability of ICT, limited availability and violation of the confidentiality and/or integrity of information stored in the ICT systems.*

Threats, Impact, Likelihood
At National level regularly the facts for cybersecurity are identified by the National Cybersecurity Centre (NCSC) This is done together by public and private parties.
The NCSC regularly drawn up the Cyber Security Assessment Netherlands (CSAN)[1] in close collaboration with public and private parties. The focus in the CSAN is the development of Cybersecurity in the Netherlands, it describes facts and is a guide based on insight and expertise of the Government.

If we have a close look at the threats of the Water management and Highways they are mostly from state actors (mostly by foreign intelligence services) or criminals by using crypto ware infections and Distribute denial-of-services (DDoS) attacks. Internal actors will become an issue as well. They have another aims, mostly financially or ideology.

Threats of the railways are different. The aim of criminals are more related to smuggling or stealing goods.

If we have a look at the most critical malicious concerns, the biggest issue of the water management is taking over the control and management systems of surface water. The interests of the railway and highways are more economic.

Sector	Manifestations	Threats: actors	Threats: tools	Resilience: vulnerabilities	Interests
Central government	The sector is often confronted with cryptoware infections and more serious and more frequent DDoS attacks. Most cryptoware is received via private e-mails sent to staff members. Spear phishing focuses on key positions within the government, for example procurement officers.	The biggest threat for the sector is posed by state actors and criminals. Internal actors are also an obstacle to cyber security. They may be motivated by financial gain, but may also have ideological motives.	Regular' criminals have easier access to powerful attack tools. Targeted attacks often take place, with information being gathered about a staff member's conduct. This information is used for a swift attack with non-digital consequences.	According to the sector, the wide dissemination of personal and other data within the government does not always entail a corresponding transfer of responsibilities for the security of such data.	The government serves a broad range of interests that depend on IT. Examples include citizens whose primary income depends on it (the Employee Insurance Agency, Social Insurance Bank), the effective actions taken by the police and judicial authorities or the control and management of surface water.
Transport (port, airport, rail)	Incidents do take place, but linking the event (for example stolen containers) to a digital attack continues to be difficult. Spear phishing attacks are sometimes highly targeted and advanced. Moreover, several crypto ware infections were detected.	Criminals pose the biggest threat to the sector. By manipulating information, they can smuggle goods or know where to steal valuable goods, for example.	According to the sector, attackers increasingly focus on process control systems. Attackers also carry out crypto ware attacks and (spear) phishing attacks.	The transport sector is characterised by its dependency on a few coordination points. Moreover, processes often become more dependent on IT due to laws and regulations in the sector. If any systems are breached, the impact of such breach will be more substantial	Short disruptions may have economic consequences as transports then take place via other countries. Long-term disruptions may lead to problems in food supply.

Table 1: Sectoral Assesment [1, appendix B]

Measures

This paragraph describes which measures should be taken by the different contracting parties.

The Dutch Government controls Cybersecurity in different ways. The "Dutch Centre for Cyber Security" (NSCS) determines the central policy on Cyber security. This policy is made by facts from public and privates parties in the Netherlands. The NSCS defines these facts from risk analyses, the threats (actor and tools) relating to cyber security.

The different sectors in the Netherlands will use these analyses of the treats to determine the right measures.
Dutch authorities such as Rijkswaterstaat, conservancies, provinces and local governments have defined their "National security Baseline information" for infrastructure where they are responsible. Based on the risk measures as established by the NSCS and the "Baseline Information Security National" prioritization of measures are adopted.
On basis of these prioritization plans are drawn up in order to carry out the selected measures. The phase to plan and implement measures called "Security by Design" by the Public Services. During the implementation of these measures it will be tested whether the desired effectiveness is achieved (can be determined that the threat level is decreasing to an acceptable level If the effectives level is not achieved, it is necessary to make adjustments. These latter phases are referred as "Security process control".

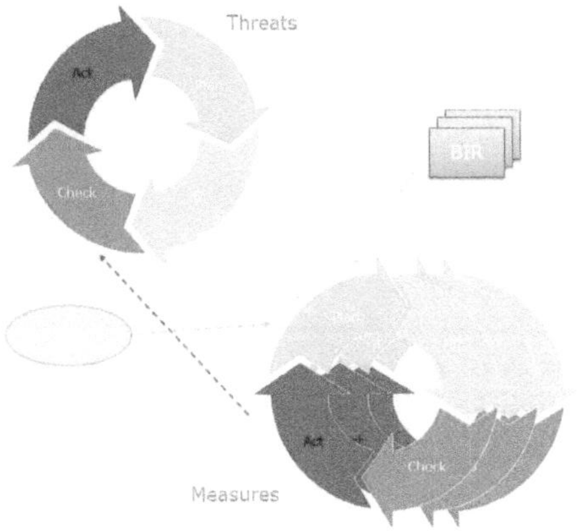

Figure 3 PDCA at different Government levels

The different governmental organisations at state, regional and local level are required to assess whether the measures taken are adequate and sorting in their effect with a minimum of a three-year cycle.

Figure 3 shows the regular process of the national government with respect to cyber threats. In addition to this, the process "Operational Security" within each of the governments is implemented. This process ensures that cyber threats occur within the regular cycle performance and that suitable action is taken. This process will be on the "Security by Design" feeding process based on event-driven risk measures to be implemented.

At National, regional and municipal levels the different Plan Do Check Act circle (PDCA) are closed and connected to each other. Therefore, information about cyber threats, measurements and effects are shared and coordinated between governments.

The following measures are examples which can be used to protect the systems

- Method defense in depth [4]
- Checklist NCSC ICS/SCADA [2,3]
- Awareness of the people, training, education

A measure for the systems in the infrastructure is using the Method defense in depth

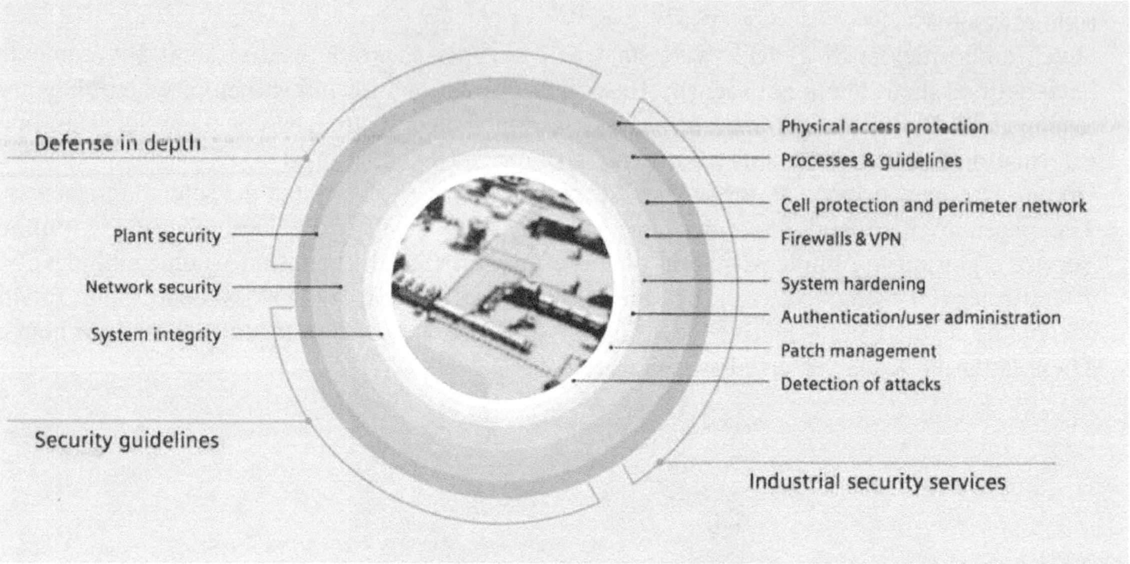

Figure 4– Method defense in Depth[4]

Summary and conclusions

In the Netherlands the awareness of cybersecurity is becoming greater, to keep the systems safe, the awareness of the people hereby is really important, the weakness can be found by the people themselves in the industrial automation.

Therefore, the awareness of the importance of System Engineering, System Thinking and the focus of holistic cyber security concerns due to the Netherlands (and other countries) system and SoS vulnerabilities to cyber-attacks/warfare. And the extensive ramifications to the countries personnel, wellbeing, etc. can be the critical success factor.

The measures are mostly focused on short terms, but it is not clear what will happen in the future. By closing the different PDCA at the different Government levels it is possible to act on changes. The focus on short term effects is in contrast with the long system life cycle of our critical infrastructure. Therefore, the Systems Engineering 'full life cycle support" concepts and concerns fill in the need for both short and long term cyber security requirements, design development/implementation, management and continuous process improvements, personnel training, etc.

References

[1] "Cyber Security Assessment Netherlands csan 2015", National Cyber Security Centrum. https://www.ncsc.nl/english/current-topics/Cyber+Security+Assessment+Netherlands

[2] "Uw ICS/SCADA- en gebouwbeheersystemen", Nationaal Cyber Security Centrum. https://www.ncsc.nl/actueel/factsheets/beveiligingsrisicos.html

[3] Checklist beveiliging van ICS/SCADA-systemen, Nationaal Cyber Security Centrum.

[4] Brochure: Security all around – Industry security for your plant – at all levels, http://www.industry.siemens.com/topics/global/en/industrial-security/pages/default.aspx

[5] "National Cyber Security Strategy", https://www.ncsc.nl/english/current-topics/national-cyber-security-strategy.html [5] Waterwet: http://wetten.overheid.nl/BWBR0025458/2016-04-14

[6] Veiligheidsnormen primaire waterkeringen
http://wetten.overheid.nl/BWBR0025458/2016-04-14#BijlageI,
http://wetten.overheid.nl/BWBR0025458/2016-04-14#BijlageII

[7] "Nationale Cybersecurity 2 – Van bewust naar bekwaam," Nationaal Cyber Security Centrum.

[8] "Achter dikke Cyberdijken – Integrale aanpak cybersecurity bij Rijkswaterstaat", RWS-iv-Partner@rws.nl

A Model Based Systems Engineering Approach to Resiliency Analysis of an Electrical Distribution Facility

Myron J. Hecht, myron.j.hecht@aero.org

Abstract

This article describes a Model Based Systems Engineering (MBSE) approach to cyberattack resilience modeling for an electrical power substation and demonstrates that (1) resiliency can be quantitatively characterized to enable design tradeoffs and (2) the analysis can be incorporated into a SysML model to enable it to be incorporated into an MBSE development effort. The resiliency analytical model described here using the Markov model, is derived from previous work on survivability analysis. For the system under analysis, it showed that the most important performance parameter for cyber resilience is propagation resistance followed by an attack rate mitigation. However, this applicability of this result is limited to this particular system with the parameter values described below. What is of more general importance is the value of such modeling, that when combined with SysML's parametric analysis capability and the differential equation solvers in tools such as Matlab, has promise in improving resiliency with respect to cyber threats in the national energy.

Keywords

SysML, Resiliency, Critical Infrastructure, Electric Power Distribution, Model Based Systems Engineering, Markov Models, IEC 61580

Introduction

Power distribution substations are critical infrastructure systems [1]. Methods emerging from reliability, availability, and probabilistic risk assessment are normally used to evaluate reliability and availability of electrical distribution systems in the presence of routine failures [2] [3]. However, such analyses are not appropriate for deliberate disruptions such as cyberattacks. This article describes Markov and MBSE modeling approaches referred to as resiliency analysis that can be used for this purpose. It also describes how such mathematical models can be encapsulated within a SysML model (which would also include a numerical solver such as Matlab) and incorporated into an MBSE process. As an example, we use a network of monitoring and control devices in an electric power distribution station.

Distribution Station and Intelligent Relay Network Description

Figure 1 shows an IEC 61850 network which consists of Intelligent Relays (also referred to as Intelligent Electronic Devices (IEDs) [4]. There are three levels: Station, Bay, and Process. At the lowest, or process level, IEDs designated as Merging Units (MUs) sample the power, current, voltage sensors as well as relay and transformer state of health data. At the bay level, Control and Protection IEDs accept input data from the MUs and processes the data. At the station level, computers provide control, maintenance and diagnostics for personnel visiting the site. However, the substation is usually unattended. Remote control signals are sent from grid operations center and travel through private and public networks to arrive at the substation and pass through the firewall and the router. Additional resiliency devices are shown on this diagram as shaded box: an external firewall, internal firewalls, and a reimaging device. Additional resiliency measures implemented in software include initial scanning, in-depth scanning.

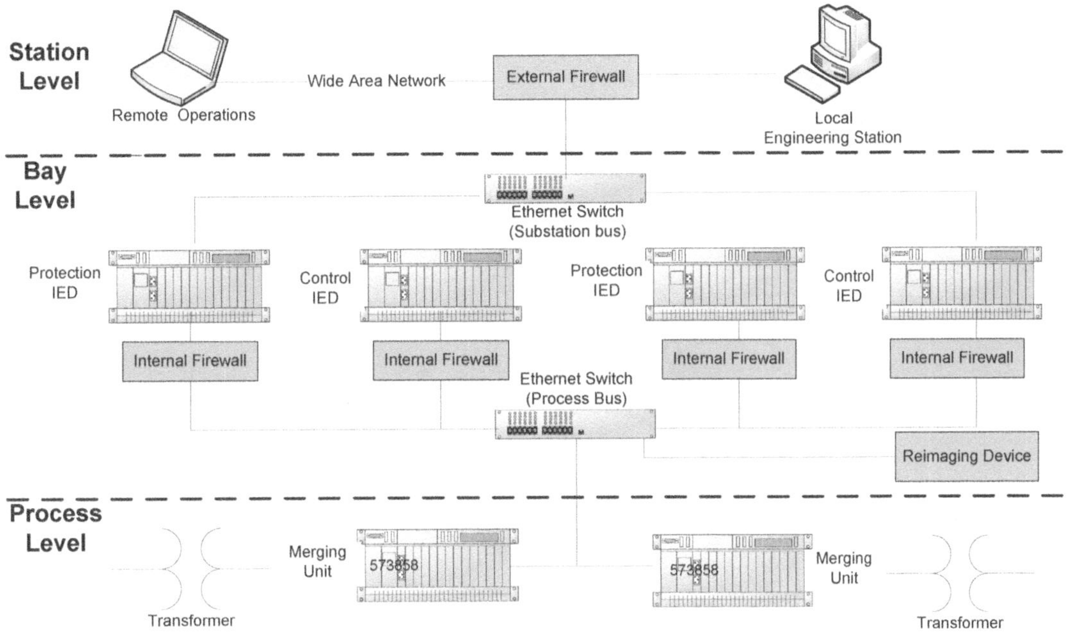

Figure 1. Electric power distribution station IEC 61850 Intelligent Electronic Device (IED) network

Resiliency Analytical Model

A resiliency analytical model enables and identifies design decisions on the network security architecture by providing quantitative evaluations. Figure 2 shows our model as an attack and restoration state transition diagram model for the network shown in Figure 1 and is derived previous work by M. Grottke, A. Avritzer, et. al. [5] [6]. This analysis treats the network as a single unit. The network starts at the initial state 0, which is the only functional state. From there, it can transition either to a real attack (state P) or a false alarm (State A). From State P, the network can transition to a short recovering state R in which the malware removed and the system rebooted or a longer recovering state Q which entails securely loading a complete trusted image from a reimaging device.

Table 1 shows the meanings of the transition rates and the mathematical expressions.

Table 2 lists the parameters in the transition expressions, their meanings, and the values. The values were taken from Grottke [6] but are not taken from measurements at distribution stations and should therefore be considered notional.

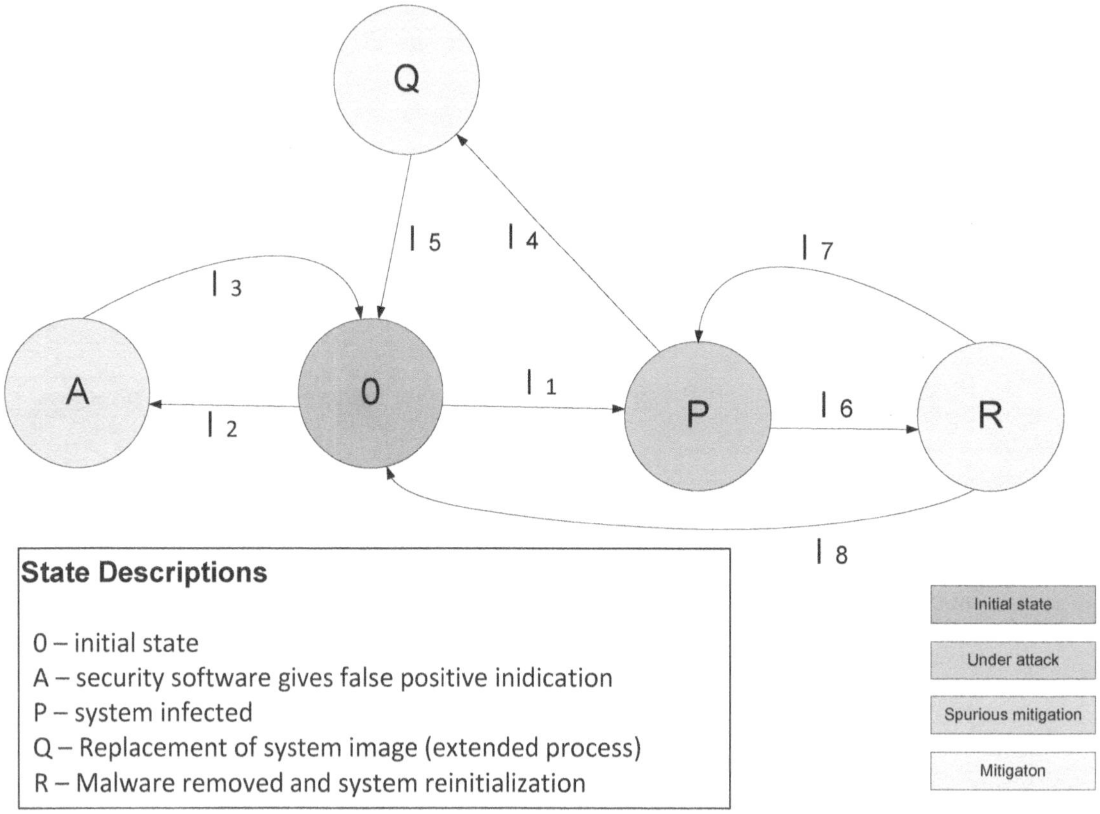

Figure 2. Resiliency Model State Transition Diagram (see Table 2 for specific expressions of each rate)

Table 1. Transitions in Resiliency Model

Transition	Meaning	Mathematical expression
λ_1	Transition to relay under attack	$\lambda_1 = v + k$
λ_2	False alarm	$\lambda_2 = R$
λ_3	Recovery from false alarm	$\lambda_3 = w*f$
λ_4	Transition severe infection (manual)	$\lambda_4 = a$
λ_5	Transition from severe infection to operations	$\lambda_5 = m$
λ_6	Transition to from actual infection	$\lambda_6 = w*d$

λ_7	Unsuccessful reload and restart	$\lambda_7 = r(1 - j)$
λ_8	Transition from reload and restart to operations	$\lambda_8 = r*j$

Table 2. Model Parameters and Associated Design Decisions

Parameter	Meaning	Baseline value	System Engineering Design Decision
v	Rate of cyberattacks	0.01/hr (Attack once every 100 hours)	Firewall to place at system boundary: greater firewall protection reduces the value of the attack rate parameter at a penalty of additional cost, complexity, and failure rates
a	rate of in depth scanning	15/hr (scanning once every 4 minutes)	Method of in-depth scanning and intrusion detection: Increasing the rate of in depth scanning reduces detection time but has a performance penalty
m	Rate of Recovery from system image replacement	0.5/hr	Amount of time necessary for initiating and reimaging a processor (requires remote initiation of reimaging and verification prior to allowing the operator to rejoin the network).
w	rate of initial scanning	30/hr (scanning once every 2 minutes)	Method of initial scanning and intrusion detection: Increasing the rate of in initial scanning reduces detection time but has a performance penalty and causes more false detections
k	rate of infection from adjacent nodes	10 (infection at a node affects 10 neighbors)	Addition of internal protections (internal firewall): additional security containment affects lifecycle cost, throughput, response time, reliability
r	Scanning completion and malware removal rate	30/hr (2 minute rejuvenation time)	Time to restore a node after an infection is affected by processor and system design
f	probability of false positive	0.3	Performance penalty of false positives
d	probability of true positive	0.7	Increasing the probability detection requires additional development and testing of intrusion detection and
j	Probability of successful rejuvenation	1	No associated design decisions. Assumed to always be 1
q	Probability of spurious rejuvenation	0	No associated design decisions. Assumed to always be 0

Table 3 shows the state flow equations. These equations express the rate of entry and exit from each of the 5 states [2]. P_x is the probability of being in state x, where x could be the states 0, A, P, Q, or R. The left hand side of equation, dP_x/dt is the likelihood of being in a state as a function of time since the beginning of the attack and is the difference between the entry and exit rates

Table 3. Rate Equations for Resiliency Model

State	Rate Equations
0	$\frac{dP_0}{dt} = -(\lambda_1 + \lambda_2)P_0 + \lambda_3 P_A + \lambda_5 P_Q + \lambda_8 P_R$
A	$\frac{dP_A}{dt} = -\lambda_3 P_A + \lambda_2 P_0$
P	$\frac{dP_P}{dt} = -(\lambda_4 + \lambda_6)P_P + \lambda_1 P_0 + \lambda_7 P_R$
Q	$\frac{dP_Q}{dt} = -\lambda_5 P_Q + \lambda_6 P_P$
R	$\frac{dP_R}{dt} = -(\lambda_8 + \lambda_7)P_R + \lambda_6 P_P$

Resiliency SysML Model Description

Figure 3 shows the two resiliency SysML model diagrams for the distribution station that were used for this analysis station (other SysML diagrams such as activity, IBD, and additional parametric diagrams would be needed for a complete description). The model was implemented in the Cameo Systems Modeler [8] SysML modeling tool.

The top diagram shows the block definition diagram consisting of a parent block, *ResilientSystem*, and are four constituent blocks corresponding to the security hardware and software installed at the substation (i.e., rollback-restoration device, initial malware scan software, intensive malware scan software, and the firewall). Two compartments are shown in the *ResilientSystem* block, the constraints and the value properties. The benefit of keeping all parameters within a model such as Figure 3 is that it provides a single repository from which engineers from any discipline can retrieve the most current values. The constraints compartment shows the name of the constraint block shown at the bottom of the table and will be discussed in the next paragraph. The value properties compartment contains all of the value properties (input parameters) shown in

Table 2. It also contains the state probabilities shown in Figure 2 as value properties.

The parametric diagram shown at the bottom of Figure 3 shows the constraint block named in the constraints compartment and the solution for the system of equations shown in Table 3. Like many other modeling tools, Cameo Systems Modeler enables parameters to be transferred directly to a Matlab function and solved using its ode45 numerical solver [9]. The blocks on the outside of the constraint block are value properties that contain the parameters listed in

Table 2. These value properties are in turn contained in the blocks associated with the design of the substation shown in the block definition diagram in the upper part of Figure 3.

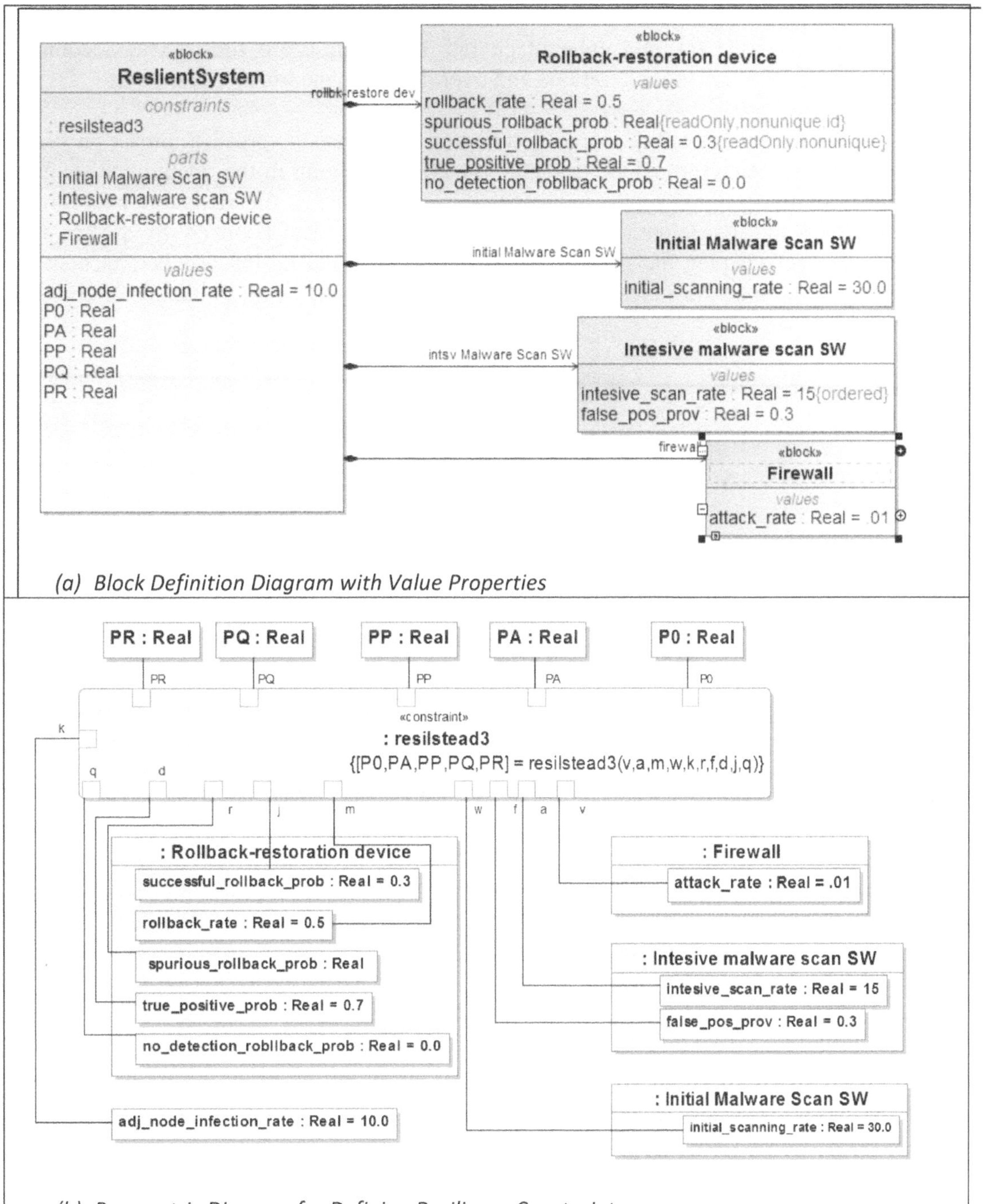

(a) Block Definition Diagram with Value Properties

(b) Parametric Diagram for Defining Resiliency Constraint

Figure 3. SysML Resiliency Model.

Analytical Results

Solutions to the Markov model described above can be either transient or steady state. The transient analysis describes the behavior of the system as a function of time. The steady state solution provides information on the long-term residence probabilities in each of the states (used primarily for availability analysis). For the purposes of the analysis described here, we are interested in the behavior of State 0, the only functional state, over the first 3 minutes (0.05

hours) of the cyberattack. We show the results of 3 parametric analyses: propagation factor (k), re-initialization rate (r), and cyberattack rate (v). These are illustrative; a complete analysis would need many more iterations and over the complete input space.

Figure 4 shows the impact of the propagation factor, k that has the greatest impact on system behavior. The ordinate (vertical axis) is the probability of being in the operational state and the abscissa (horizontal axis) axis represents time. Increasing the defenses between the intelligent relays through internal firewalls or restricted physical connectivity are the most effective means of keeping the distribution station operational during a cyberattack in this model. Of course, once the attack is stopped, the probability of being in the operational state goes back to 1.

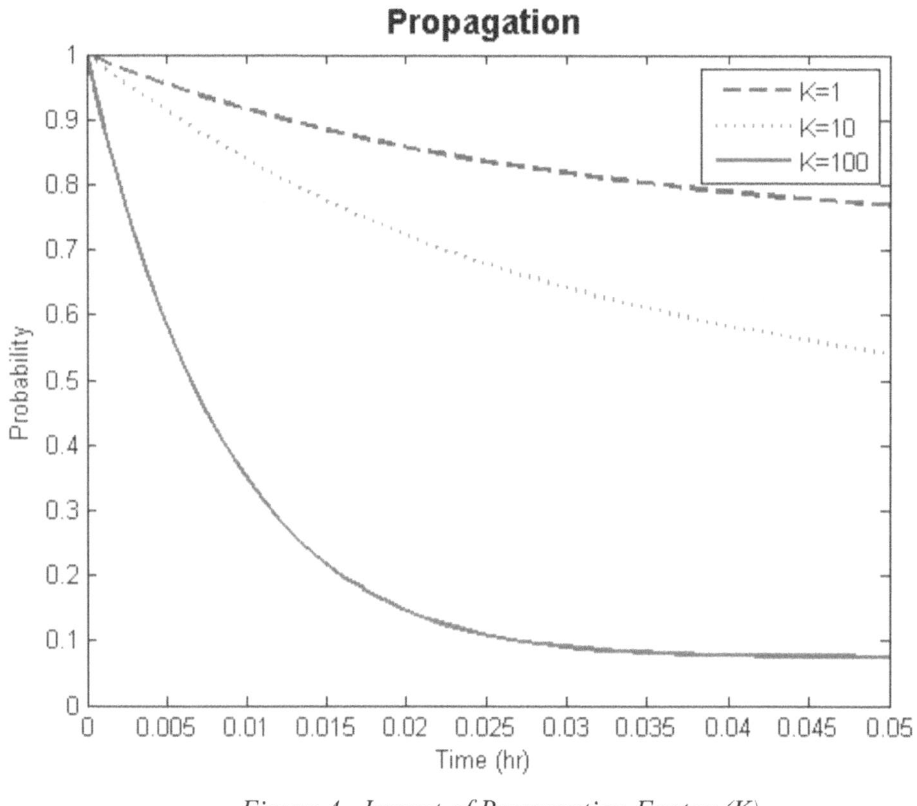

Figure 4. Impact of Propagation Factor (K)

Figure 5 shows that the impact of the first level re-initialization rate is less than the propagation factor (the axes have the same meaning as in Figure 6). The Re-initialization (state R) is the first line resiliency defense, and a higher rate has a greater benefit than the processor re-imaging and recovery rate (from state Q). When the attack stops, the system will return to State 0.

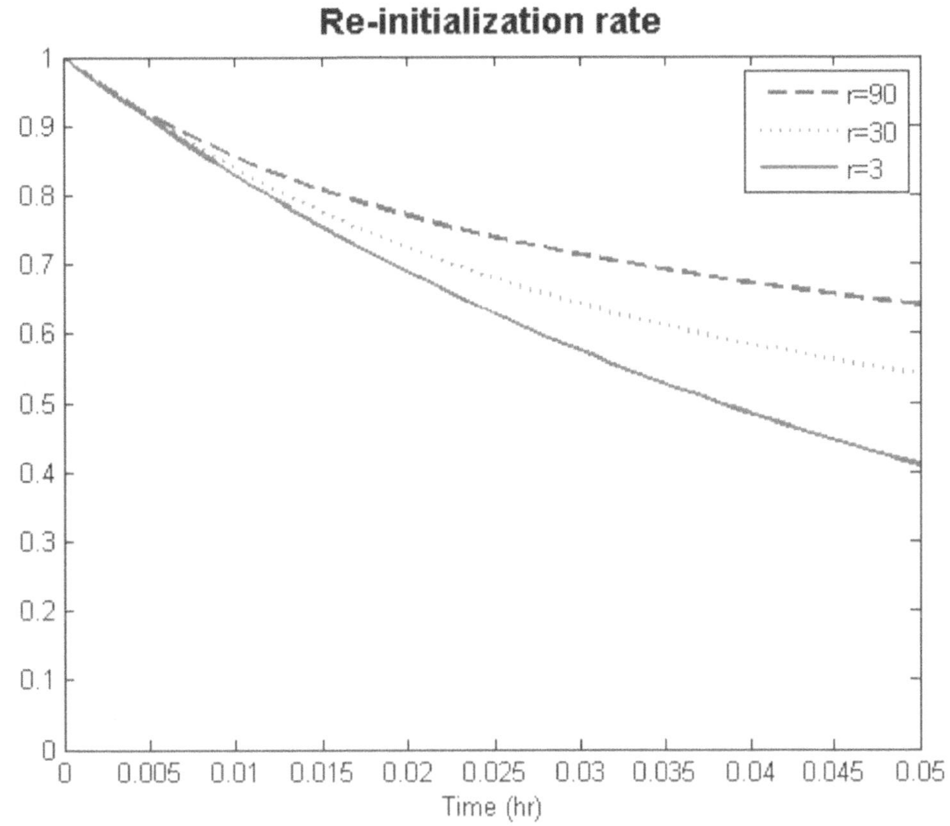

Figure 5. Impact of Re-initialization Rate (r)

Figure 6 shows the impact of the successful cyberattack rate. At these rates, (between 1 per hour and 0.01 per hour), the impact is not high because the recovery rates (either re-initialization or re-imaging) are high relative to the attack rate. However, if the attack rate were increased to 100 per hour, then the impact would be much higher. This result shows the importance of firewalls and other defenses to attacks. The more effective such defenses, the less frequent the transitions to an infected or recovery state and the higher the overall availability.

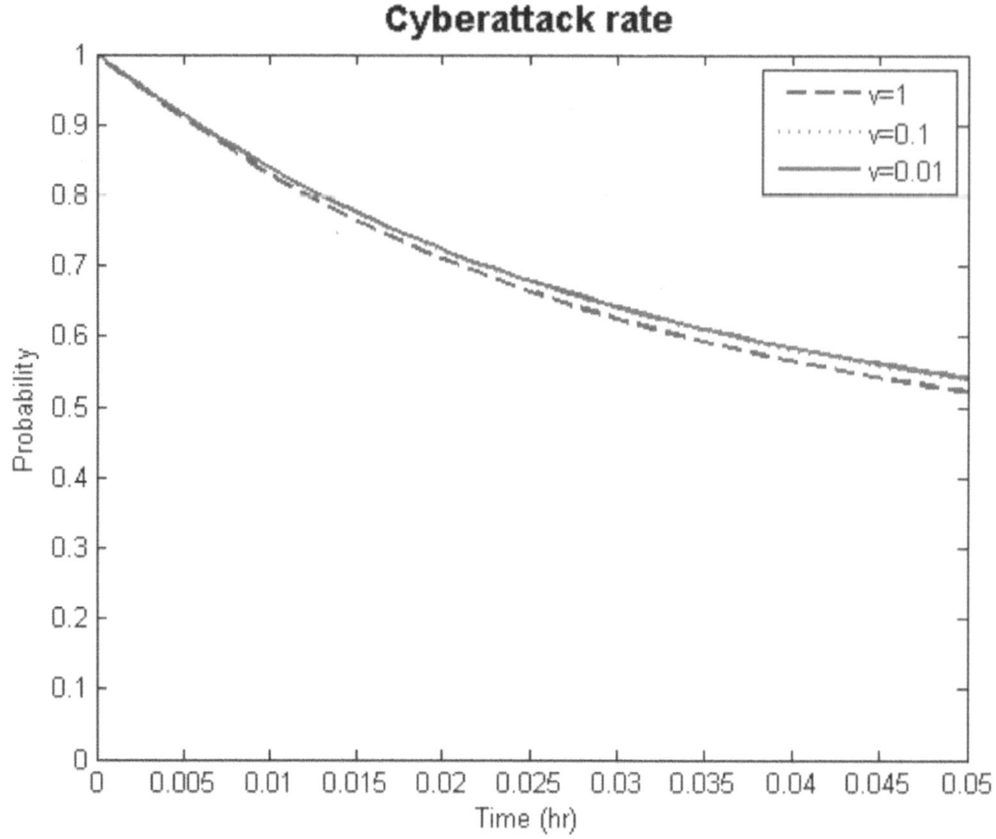

Figure 6. Impact of Cyberattack Rate (v)

The model results show that preventive mechanisms such as propagation resistance (containment) provide a greater benefit (in terms of the probability of staying in the operational state, see Figure 4) than increasing recovery rates of mechanisms such as re-initialization or re-imaging (see

Figure 5 and Figure 6). However, in the absence of recovery mechanisms, there would be no return paths to the operational state after an attack. Therefore, such restorative mechanisms are necessary to prevent the distribution station from becoming inoperative.

Conclusions

This article showed how SysML models of electrical distribution systems can incorporate resiliency analyses. In addition, other analyses for reliability, maintainability, acquisition cost, and sustainment cost can also be included in SysML. This approach shown here is an example; it is not yet part of standard design practices at any utility. However, when such projects begin using Model Based System Engineering, all of these values will be kept within the model. Because the analytical model is integrated with the rest of the system model, the analysis can be readily repeated at any time and the results can immediately be made available to the system designers for consideration and if necessary iteration. The immediate feedback of the analysis provides more rapid feedback and lower cost analyses that in turn facilitates optimum and rapid resilient designs using MBSE and parametric analysis.

References

[1] Department of Homeland. Security, "What Is Critical Infrastructure?," 16 January 2016. [Online]. Available: https://www.dhs.gov/what-critical-infrastructure. [Accessed 11 March 2016].

[2] J. Zhang and C. A. Gunther, "IEC 61850 - Communication Networks and, Systems in Substations : An Overview of Computer Science, University of Illinois at Urbana Champaign," 2011. [Online]. Available: http://seclab.web.cs.illinois.edu/wp-content/uploads/2011/03/iec61850-intro.pdf.

[3] F.M. Alwan A. Baharum, and G.S. Hassan, "Reliability Measurement for Mixed Mode Failures of 33/11 Kilovolt Electric Power Distribution Stations," *PLoS ONE,* vol. 8, no. 8, 2013.

[4] A. Baharum, F.M. Alwan, and S.T.Hasson, "A Case Study of Reliability and Performance of the Electric Power Distribution Station Based on Time between Failures," *Mathematical Problems in Engineering,* vol. 2013, no. 583684, p. 6, 2013.

[5] A. Avritzer, L.Carnevali, L. Happe, A. Koziolek, D. S.Menasché, M. Paolieri and S. Suresh," A Scalable Approach to the Assessment of Storm Impact in Distributed Automation Power Grids in *Quantitative Evaluation of Systems (QEST)*, Florence, 2014.

[6] M. Grottke, A. Avritzer, D. S. Menasche, J. Alonso, L. Aguiar, and S. G. Alvarez, "WAP: Models and Metrics for the Assessment of Critical-Infrastructure-Targeted Malware Campaigns," in *International Symposium on Software Reliability Engineering*, Gaithersburg, MD, 2015.

[7] C. Ebeling, Introduction to Reliabiltiy and Maintainability Engineering, 2nd Edition, Illinois: Waveland Press, 2010.

[8] NoMagic Corporation, *Cameo Systems Modeler version 18.3,* http://nomagic.com/products/cameo-systems-modeler.html, 2015.

[9] Mathworks Corportation, ode45 differential equation solver description, http://www.mathworks.com/help/matlab/ref/ode45.html

How Infrastructure Can Become Reborn by Becoming Born Robust for the Critical Infrastructure Protection and Recovery Working Group

Josh Sparber, jsparbear5@gmail.com

Abstract

SysML is a tool for guiding engineers in designing power grid circuits sufficiently robust to withstand known Electromagnetic Pulses (EMPs). Careful examination of existing data shows that EMPs, and sometimes Geomagnetically Induced Currents (GICs) that accompany them are truly a powerful threat to power grid survival. Systems engineers, employing System Modeling Language (SysML) can isolate power grid failure susceptibilities and areas for necessary power grid design improvements with selected SysML packages defined as enclaves associated with risk. These enclaves can be decomposable into stereotyped components available for risk categorization, building simulation libraries, or follow–on test. As an example, a stereotype Source, instantiated as a Photovoltaic (PV) Inverter, increasingly important in microgrid renewable energy, is linked to a High Frequency AC (HFAC) microgrid risk enclave package. Simulation allows evaluation of SysML Use Cases with EMP Actors. Real world test, construction, and strategic grid readjustment can then segue quickly.

Keywords: Electromagnetic Pulses, Microgrid, Photovoltaic, Satellite, Simulation, SysML, Use Cases

Introduction: Constructing the Robust Power Grid

Why This Paper Was Written

This article illustrates how using risk based profiles can build bottom up constructions of structures resistant to EMPs within a microgrid. With an assemblage of components tested for risk, simulated as SysML EMP threat Use Cases, an engineer can test and redesign selected portions of the power grid against EMP vulnerability and find paths to enhancing survivability or improving the reliability of the enclosing system.

The Power Law Nature of Power Outage Events

The Failure of Risk Management describes how a power outage is in a class of events that best follow the power law distribution rather than the normal distribution (Hubbard 2009, 184). The class members are "volcanic eruptions, forest fires, earthquakes, *power outages* [emphasis mine], asteroid impacts, and pandemic viruses" (Hubbard 2009, 184). These events can seriously affect "stressed systems that allow for both common mode failures and [cascade of failures]" (Hubbard 2009, 185). Increasing complexity now exposes the grid to new vulnerabilities. EMP power outages have and probably will cause high–intensity failures more regularly now. EMP events have the strong potential to cause a more severe impact on the power grid than expected.

The Study of Solar Events

Solar flares and Coronal Mass Ejections (CMEs) cause radio blackouts, radiation storms, and geomagnetic storms as effects (US National Oceanic and Atmospheric Administration. n.d.). Solar events can severely impact grid structures, as solar flares striking these earthbound structures or Geomagnetically Induced Currents (GICs) from CMEs burning through these structures. Intensive Kepler satellite data shows that stars like our sun, G main sequence stars, are capable of flares much larger than those previously recorded (Shibata 2016). According to data taken from 80,000 stars, even "superflares," flares 100 to 1,000 times anything previously recorded, can occur in sun–like stars as frequently as every 800 to 5,000 years (Shibata 2016).

Since World War II, solar dynamics are known to contribute to Earth weather. The science of helioseismology enabled scientists to study the internal solar "architecture" that leads to solar effects (Morrow 2015). While many satellites now fielded (SOHO, STEREO, SDO) observe both the nature of the sun and similar stars for gathering data for more effective predictions, scientists have a poor understanding of Space Weather dynamics and cannot predict it reliably (Thompson 2015).

An insufficiently prepared grid system could collapse even due to a relatively minor weather event, intentional mischief, or even decay or neglect. Increasing human and natural threats means that the probability of an EMP event causing damage may happen with greater intensity and greater regularity than in previous times. It makes sense to examine fortifications that will efficiently protect the existing power infrastructure. How can we build a power grid system that will provide this protection?

Building a Model: Analysis

In the following paragraphs, the author will discuss an approach to organizing portions of the power grid for understanding risk through a model–based system model, and simulated for test based on that system model or an improved "should be" design with enough robustness to withstand impacting EMPs or GICs. The depicted power grid system is an HFAC microgrid. The author intentionally chose this grid utility—it is a hybrid between a microgrid fitted to conventional AC power and the currently popular and nonconventional DC microgrid. A solar voltage source serves as an example of a vulnerable component. Microgrids and the renewable energy of solar power will be having greater and greater impact in constructing a robust grid in the coming years.

Starting with Block Diagrams

Friedenthal, Moore, and Steiner 2009 state that a generic type for a part can be "modeled in SysML with a block definition diagram, bdd" (34). While a bdd usage can be for a generic functional unit, specific standard functional units derive from the generic functional unit. The use of these derived block diagrams, called internal block diagrams (ibds) (Friedenthal, Moore and Steiner 2009, 44), will be demonstrated later.

Stereotypes and Block Definition Diagrams

According to Friedenthal, Moore, and Steiner 2009, "Stereotypes are used . . . to customize the language for specific domains" (542). For example, functionally distinct portions of microgrids could be framed as microgrid bdd stereotypes: sources, transducers, converters and inverters, protectors, storage/recovery devices, communication devices, assets, interfaces, and the connected main grid power components. There is no display of management and control elements—only a few sample functional breakdowns.

Table 1. *A few microgrid component bdd stereotypes and possible specific ibds*

bdd stereotype	device ibds	bdd stereotype	device ibds
Main Grid Components	Power Transformers, VVR–VVO circuits, High Power Caps	Sources	Solar, Wind, Small Hydro, Diesel, Geothermal, Co–Generators
Converters/Inverters	AC/AC, AC/DC, DC/AC, DC/DC	Protectors	Fuses, Ground Fault Detectors, Spark Gaps, Circuit Breakers
Storage/Recovery	Ultra/Supercapacitor, Flywheel/Battery Farms, Compressed Air Energy Storage, Hydrogen Storage, or Pumped Hydro	Communication	Risk Applications, SCADA, Telemetry, GPS

Internal Block Diagrams

For recasting power infrastructure into a more robust system, SysML "provides a mechanism for dealing with legacy system elements that have not been developed using rigorous modeling techniques"; i.e., a 'profile' (Friedenthal, Moore, and Steiner 2009, 307). This author is interested in building profiles within grid portions specifically oriented to risk. Groups of block definition diagrams would characterize a particular enclave or supergroups of bdds associated with particular types of risk information. Once fully characterized, this profile information could be passed on to any grid portions of similar composition for both risk classification and risk comparison.

Ibds would be the actual realizable components for simulation and later test, within those bdds. Ibds would convert into SysML State Machine diagrams for checking changes to internal parameters under EMP or GIC stresses. Ibds within this discussion are blocks within other blocks; however, most fully developed ibds show interconnections within and between each other and to other external bdds.

Look at the inverter below. The Table 1 bdd stereotype 'Sources' could be the bdd for the instantiation of a solar source ibd as the PV inverter in Figure 1. Another ibd instantiation could be a Concentrated Solar Power receiver, another type of solar energy source. The American solar industry planned building 7 Gigawatts of solar stations of over 1 Megawatt capacity in 2016 (Greentech Media 2015). PV inverters will play an important role. In Figure 4, the Solar Source ibd is a 'part' within the Sources Stereotype block.

Figure 1. *PV Inverter Circuit Source (Next Electronics 2016)*

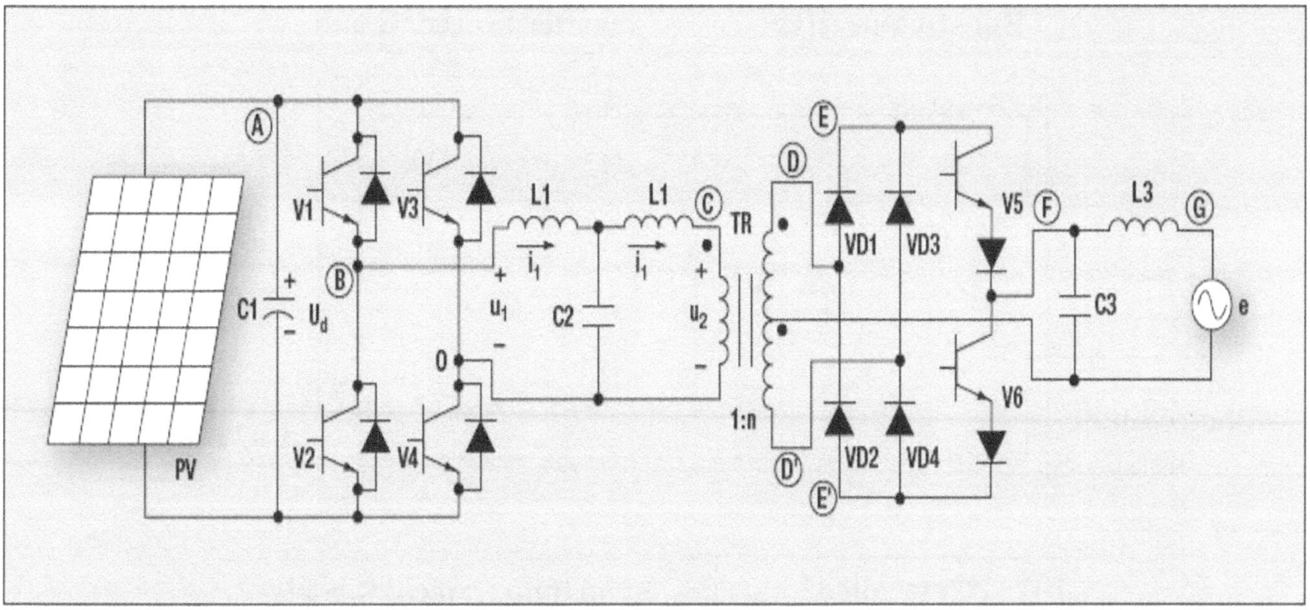

Fig. 1. *System topology of the proposed single-phase current-source grid-connected inverter that consists of a high-frequency, full-bridge inverter, immittance converter, center-tapped transformer, high-frequency bridge rectifier, power-frequency inverter and low-pass filter.*

Use Cases and Simulation

Manmade Nuclear High Altitude Electromagnetic Pulses (HEMP) contain E1 (extremely short duration), E2 (moderate duration), and E3 (long duration) portions (Pry 2013, 5). E1 and E3 are the gravest danger to equipment (Pry 2013, 5). Naturally occurring EMPs; e.g., CMEs, have effects similar to E3; that is, they give rise to powerful GICs that travel through the ground—there is a strong claim for emphasizing testing power transformer neutrals to GIC exposure (Emanuelson 2013a, 5). Conspicuous interconnection length between parts or an interconnection length subject to harmonic resonance is vulnerable to E1 type pulses (Emanuelson 2013b). A simulation can highlight where circuit vulnerabilities exist to either an EMP pulse or a GIC moving through the ground, connectors, or vulnerable components (Hathaway and Byers 2015).

There are many ways of preparing a PV inverter circuit for simulation. Simulink can be used to build a simulation model of the PV inverter by connecting subcomponents (Osorio 2014). Full detail modeling of system parameters: voltage, current, thermal, or power properties, can also be done in a "Sim Power Systems" application (Osorio 2014), or in Matlab, see Figure 2.

Figure 2. *PV inverter source functional blocks leading to a detailed Matlab simulation (Mollah, Panda, and Saha 2016, Figure 1)*

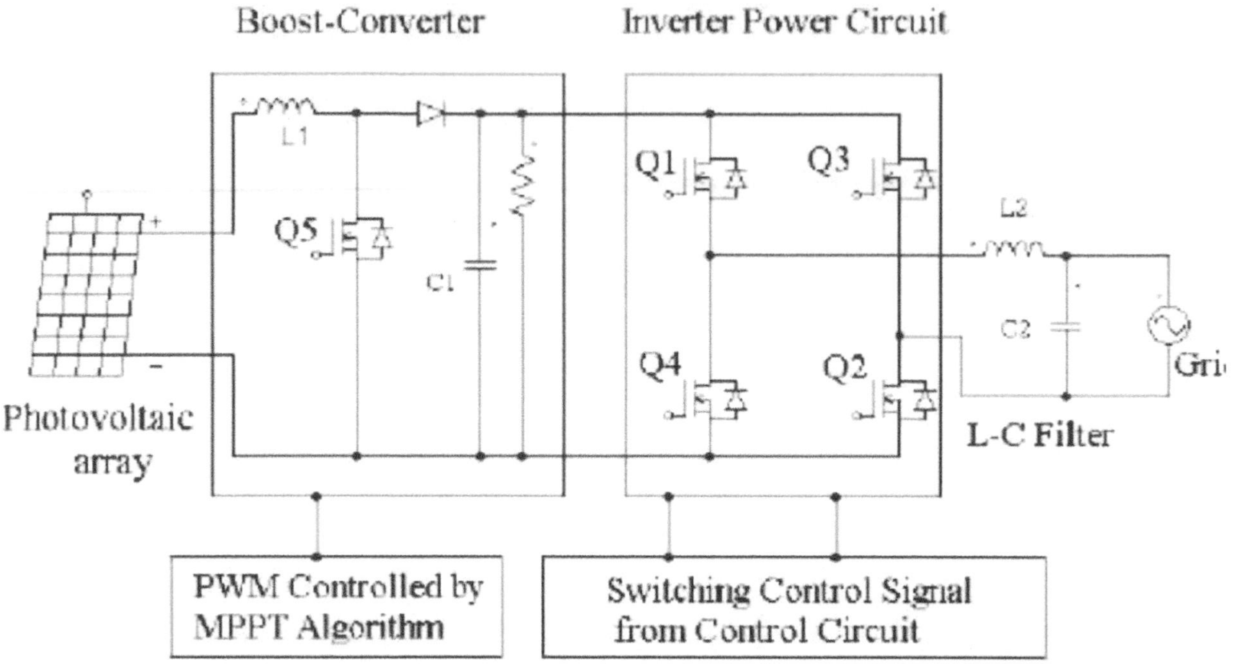

Figure 3 shows Figure 2 as a possible ibd precursor to a State Machine diagram ready for simulation. PTC Integrity Modeler 8.2 allows sufficiently detailed lower level Use Cases to be connected to simulations by allocating activities to State Machine diagrams or adding 'Constraint Blocks' to ibds (PTC 2013b, 77–79, 86–89). In PTC Integrity Modeler 8.2, a 'SySim Profile' can also apply to an enclave of "at risk" components in preparation for simulation of a detailed lower level Use Case (PTC 2016a, 3). In PTC Integrity Modeler 8.2, ibds 'owned' by risk enclaves, with a 'SySim ControlBlock Stereotype' assigned, are ready for front–end analyses runnable in Visual Studio (PTC 2016a, 4, 11). Access to Simulink is also available in PTC Integrity Modeler 8.2 (PTC 2016a, 3).

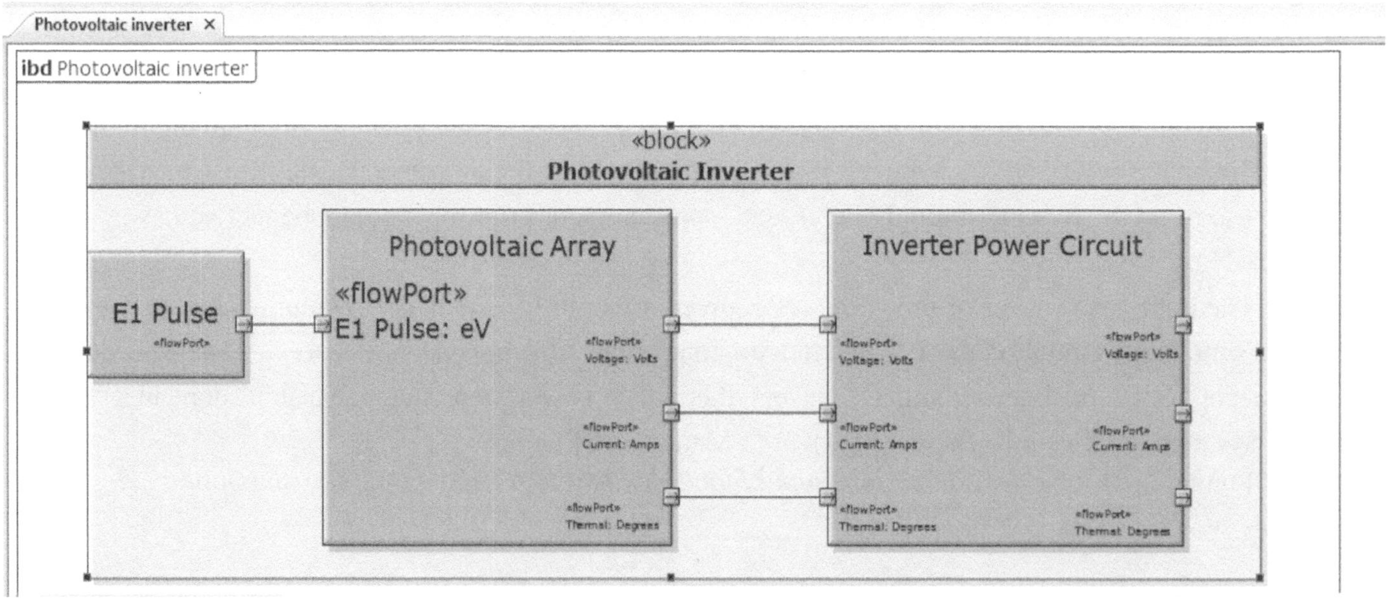

Figure 3. *Ibd with state inputs shown*

Building a Model: Synthesis

"A package is a container for other model elements" (Friedenthal, Moore, and Steiner 2009, 81). In this discourse, generic functions are stereotyped bdds. Let packages contain these stereotyped bdds in an enclave of "at risk" components in a "risk profile." These enclaves would be microgrid portions available for EMP/GIC impact evaluation. Figure 4 is an HFAC microgrid. Create a stereotyped Sources bdd for Solar Power and Wind Power inputs. Then, finally instantiate the Solar Power bdd as an ibd of the PV Inverter circled below in Figure 4. The Main Grid component stereotype bdd could contain the ibds for the AC Line, Linear Load, Motor, and the Distributed Network. The Converter/Inverter stereotype could contain the AC/AC converter ibd.

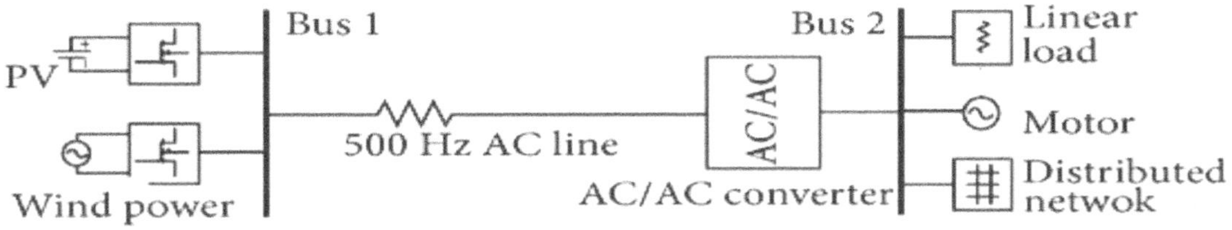

Figure 4. *HFAC Microgrid (Mariam, Basu, and Conlon 2013, Figure 6)*

A risk package enclave for a microgrid under review could ultimately contain the whole set of stereotyped bdds. See Figure 5. A larger package could form a "risk profile" of such components to be tested for robustness and reintegration that could be used to characterize all similar microgrid portions. Friedenthal, Moore, and Steiner 2009 state, "A profile is a kind of package used as a container for a set of stereotypes and supporting definitions" (346).

A microgrid level risk profile could contain all the bdds, Use Cases for the simulation, and also any standards that would apply. See figure 6. In PTC Integrity Modeler 8.2, a "risk profile" could be customizable as an 'ergonomic profile' (PTC 2016b). To create such a custom profile takes "a good working knowledge of the VBScript language, the Modeler Meta Model and the Modeler Automation Interface" (PTC 2016b).

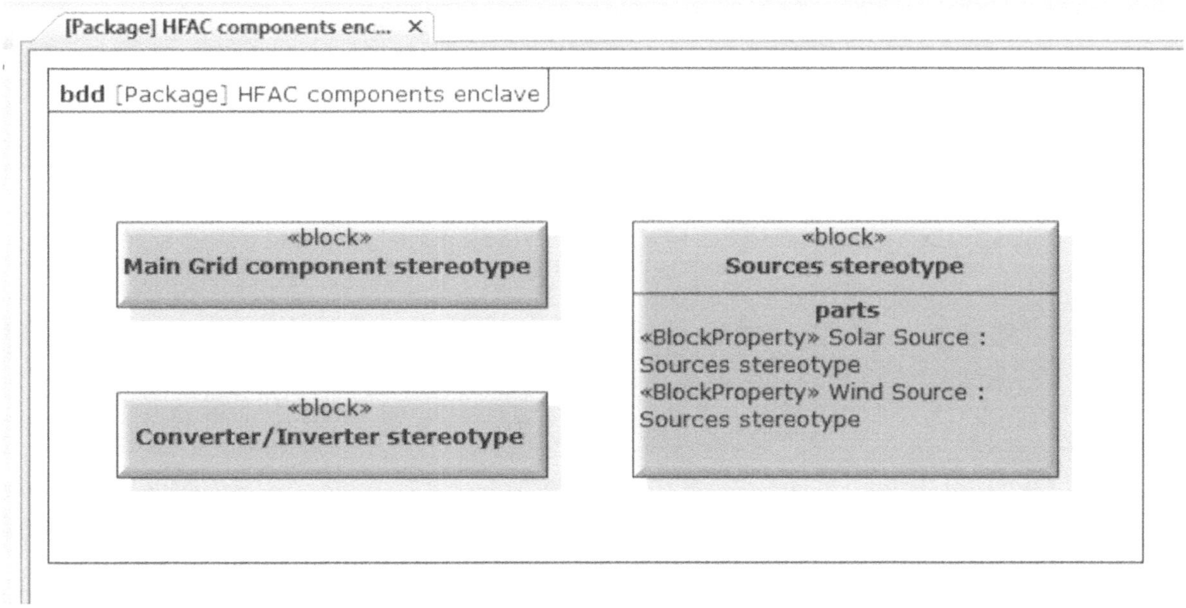

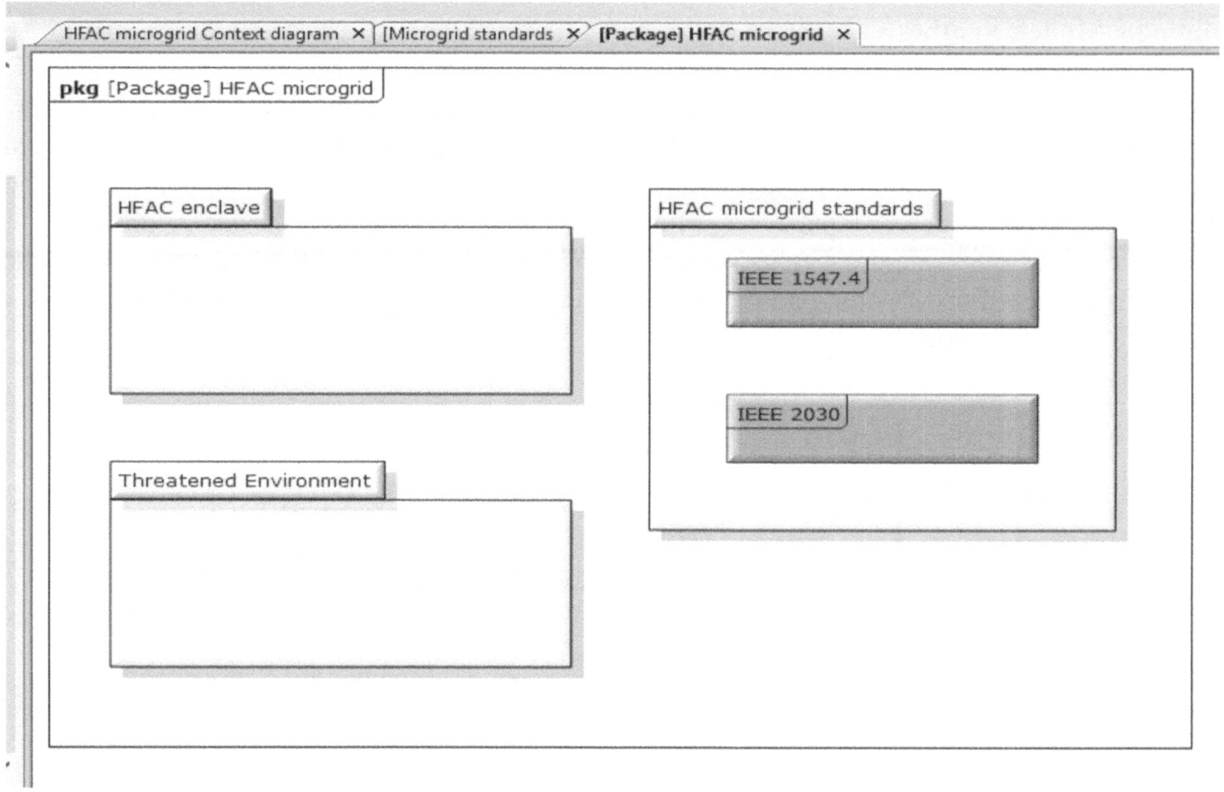

Figure 5 and Figure 6. *Packages for HFAC microgrid (PTC 2016b; Etemadi 2013, 22; Basso 2014)*

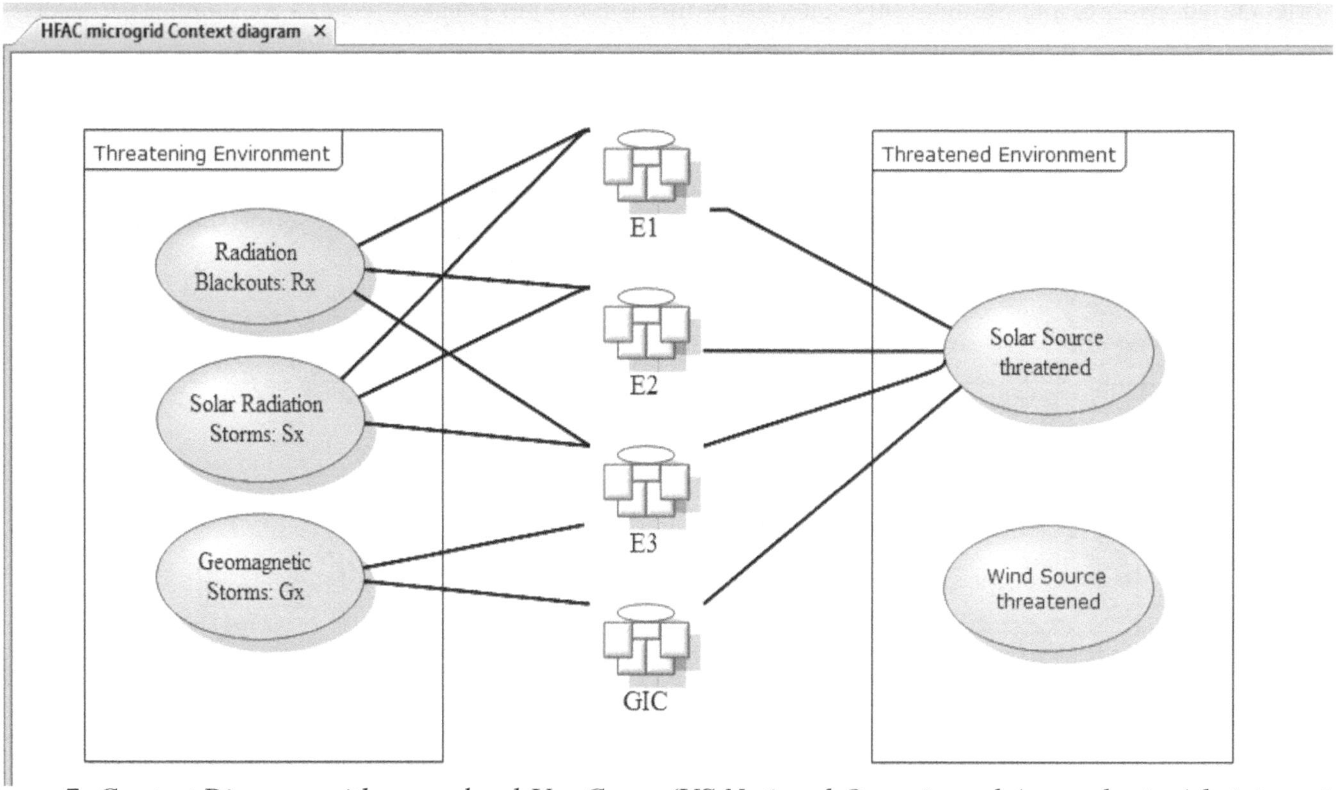

Figure 7. *Context Diagram with upper level Use Cases (US National Oceanic and Atmospheric Administration. n.d.). N.B.: HFAC Threatened Environment Use Cases are referenced in the HFAC microgrid package shown in Figure 6.*

Conclusion

To this author's mind, Model–Based System Engineering can achieve a selective, strategic deployment of architectures composed of robust elements through SysML. The use of groups of risk profiles, common to many power grid architectures; e.g., microgrids, could determine categories of components needed to be systematically simulated, prepared for test, and for possible later redesign and follow–on retest. Stereotyped components would help classify exactly what portions of a particular grid structure had needs to address vulnerabilities to various EMPs or GICS. A vast library of common simulations could be aligned to risk classifications available to any engineer.

Power grid structures, recombined with robust components, will possess high survivability after thorough test and needed redesign. Constructions based on intensively tested components would be field ready for incorporation into any portion of the power grid needing survivability enhancement.

Simulation will be key to facilitating more robust designs. Any new constructions will have a need for thorough test. Simulation allows for a great variety of architectures to be quickly and thoroughly tested before anyone commits to a Bill of Materials or a single rivet goes into a support structure.

References

Basso, T. 2014. "IEEE 1547 and 2030 Standards for Distributed Energy Resources Interconnection and Interoperability with the Electricity Grid." NREL. Accessed Sept. 3, 2016. http://www.nrel.gov/docs/fy15osti/63157.pdf.

Emanuelson, J. 2013a. "E1, E2, and E3." Accessed May 11, 2016. http://www.futurescience.com/emp/e1-e2-e3.html.

———, J. 2013b. "EMP Myths." Accessed May 12, 2016. http://www.futurescience.com/emp/EMP-myths.html.

Etemadi, A. 2014. *Microgrids: Operation, Control, and Protection.* Saarbrüken, Germany: Lambert Academic Publishing.

Friedenthal, S., A. Moore, and R. Steiner. 2009. *A Practical Guide to SysML: The Systems Modeling Language*. Burlington, US–Massachusetts: Morgan Kaufmann OMG Press.

Hathaway, K. and B. Byers. 2015. "LT Spice IV: Building and Simulating a Buck Converter," video. Accessed Mar. 5, 2016. YouTube. https://www.youtube.com/watch?v=pGo8XlZ7U4s&ebc=ANyPxKomm47hnKPXAxtkZdmWkjwqMrGy6fgToYj8abbVcaYB7_hhaGhB2EZPeoxIWp6z6TGMbS7oSNYI6q4kwMRfiP9tgvj41A.

Mariam, L., M. Basu, and M. F. Conlon. 2013. "A Review of Existing Microgrid Architectures." *Journal of Engineering* 2013, no. 637614: 1–10. Dublin Institute of Technology. Accessed Mar. 6, 2016. http://dx.doi.org/10.1155/2013/937614.

Mollah, A. H., G. K. Panda, and P. K. Saha. 2016. "Single Phase Grid Connected Inverter for Photovoltaic System with Maximum Power Point Tracking." *International Journal of Advanced Research in Electrical, Electronics and Instrumentation Energy.* Accessed Sept. 9, 2016. http://www.rroij.com/open-access/single-phase-gridconnected-inverter-forphotovoltaic-system-with-maximum-powerpoint-tracking.php?aid=43213.

Morrow, A. 2015. "A Look Back at NASA Solar Missions." US NASA Goddard Space Flight Center. Last updated Dec. 1, 2015. Accessed Feb. 16, 2016. http://www.nasa.gov/feature/goddard/nasas-soho-celebrates-20-years-of-spacc-based-science.

Next Electronics. 2016. Accessed Feb. 23, 2016. http://www.next.gr/uploads/34/807PET21-photovoltaic-systems-Figure01.jpg.

Osorio, C. 2014. "Modeling and Simulation of PV Solar Power Inverters," (video). YouTube. Accessed Apr 21, 2016. https://www.youtube.com/watch?v=GnZFi9CzF9Q.

Pry, P. V., Dr. 2013. "Electric Armageddon." Createspace: World Wide Web.

PTC. 2016a. PTC Integrity Modeler SySim Tutorial. Version 8.2. Accessed July 7, 2016. http://www.ptc.com/model-based-systems-engineering/integrity-modeler.

——— 2016b. PTC Model–Based Systems Engineering Tutorial. Version 8.2. Accessed July 7, 2016. http://www.ptc.com/model-based-systems-engineering/integrity-modeler.

Shibata, Kazunari. 2016. "Superflares on Solar type Stars and Their Implications on the Possibility of Superflares on the Sun." Lecture at the Space Weather Workshop, Broomfield, Colorado, US, Apr. 25–29. Accessed Oct. 4, 2016. www.swpc.noaa.gov/sites/default/files/images/u33/final_shibata_SWW_2015.pdf.

Thompson, M. J. 2014. "Grand Challenges in the Physics of the Sun and Sun–Like Stars." *Frontiers in Astronomy and Space Sciences* (June 26, 2014). Accessed Feb. 16, 2016. http://dx.doi.org/10.3389/fspas.2014.00001.

US National Oceanic and Atmospheric Administration. n.d. "NOAA Space Weather Scales." Space Weather Prediction Center. Accessed Sept. 9, 2016. http://www.swpc.noaa.gov/noaa-scales-explanation.

About the Author

Josh Sparber, receiving air radar training in the USMC 1975, spent 20 years in the electronics industry, 1980–2000. After earning an MSEE from Cal State Fullerton 1999, he familiarized himself with the practice of System Engineering for 16 years with the US Department of Defense. He has been an INCOSE member since 2005 and received his Certified System Engineering Professional certification in 2007. He has also persisted in a lifelong interest in environmental issues. He received a Master's Degree in Environmental Policy and Management from the University of Denver in 2015, in which his thesis focus was the search for possible system solutions to rationally and robustly reconstruct the US National Power Grid against the impact of naturally occurring Electromagnetic Pulses. The focus on protecting the power grid from natural EMPs is only one small piece of a very large solution, but it helps energize the conversation on how to protect, grow, or extend our existing infrastructure. The model Mr. Sparber is currently building in PTC Integrity Modeler 8.2 he calls Risk Exploration Verification Integration Validation and Evaluation, REVIVE.

Software and System Integrity Assessment

A Comprehensive, Stop-loss Opportunity for Critical Infrastructures

Jack Ring, jack@ontopilot.com

Abstract

This article describes how the digital aspects of all critical infrastructures world-wide can be evolved faster and made less vulnerable to cyber actions. The purpose is to motivate systems practitioners to become adept at software and system integrity assessment. We introduce the theory and method for discovering all logic, arithmetic and semantic faults throughout an arbitrarily large, heterogeneous software suite. Also, we highlight a way of automating integrity assessment. One approach was initiated in the 1990's, proven useful by the Peritus Corp. then patented in 2002. With the advent of OntoPilot LLC in 2008 a simpler and more effective approach has evolved the essentials of which are ready for release in the public domain. The balance of this article introduces the situation, highlights the system principles involved, considers the challenges in assessing software integrity, highlights the opportunity and value proposition, outlines the automation technology and finishes with a call to action. We recommend that certification in system integrity assessment be considered as a facet of the professional certifications offered by INCOSE, NDIA, IISE, and other professional and industry associations.

1. Situation

The U.S. Presidential Policy Directive, Critical Infrastructure Security and Resilience/PPD 21 states objectives such as **strengthen**, **reduce**, **minimize** and **hasten.** These are appropriate but do not go far enough. For cybersecure systems we must be more ambitious and specific. We must **commit** to specific goals such as Zero Vulnerabilities and Zero Loss, **create** the technology, tools and talent base to meet these specific goals, and **act** vigorously until ---

a. already deployed systems in all 16 critical infrastructure categories are cleansed of latent vulnerabilities.
b. all four kinds of on-going maintenance changes acknowledged in ISO/IEC 14764:2006 (corrective, preventive, adaptive, and perfective) do not result in any new vulnerabilities.
c. all development of replacement and additional systems throughout the 16 critical infrastructure categories are free of deployed vulnerabilities.
d. the 17th and most critical infrastructure, notably the sociotechnical system we all must become in order to accomplish all this, is also operational and fault-free. The 'we' here must involve government, industry and academia and go beyond collaboration to co-learning.

2. System Principles

Two system principles are key to achieving these objectives. A system has dynamic limits and integrity limits. These determine the envelope of system effectiveness in any given usage scenario.

Dynamic limits constrain system responses to stimuli. These limit the variety to which the system can respond, both semiotic variety, e.g., what kinds of stimuli and temporal variety e.g., how fast and how long.

Integrity limits clarify the conditions under which a system cannot produce appropriate responses to authorized stimuli or cannot preclude responses to unauthorized stimuli. Beyond its integrity limits the resulting configuration is simply not capable of accomplishing the intended effects.

3. Integrity Assessment Challenges

The dynamic limits of a Critical Infrastructure system a) can be anticipated by analysis or simulation of the system model provided we take care to ensure that all key factors are included. [Friedman, G 2005, Constraint Theory: Multidimensional Mathematical Model Management, Springer.] and/or b) can be discovered when exercising the actual system in a pseudo-operational environment as when testing the system and/or in an actual operational environment. The findings from exercises must be analyzed to discern Truth, False Positives and False Negatives.

Integrity limits reveal a) incoherent design especially among multiple system elements, b) shortfalls in system resources, c) hardware failures, d) latent software faults, e) inappropriate maintenance actions and f) vulnerability to unauthorized configuration changes, also called cyber actions, whether caused by things external to the system or things and actions within the system.

Inherent integrity limits can be discerned by inspecting the system design materials and proofreading the code as text. We call this Software Integrity Assessment, SIA. Other integrity limits can be discovered only when they happen while the system is being operated. Examples are hardware failures, improper data base administration actions, improper maintenance changes and security attacks. These are revealed when the system 'hangs,' aborts, outputs error messages or produces obviously erroneous results.

A key challenge is for the assessment capability to be present when these failures occur. This means that assessment is not just something done one time during the system development phase. Instead an adequate assessment capability must be built into every system because even though a deployed system operated fine the last time the operator cannot assume it will do so next time [Key Challenges and Opportunities in 'System of Systems' Engineering, 2005, J. Ring and A. Madni, Proceedings of 2005 IEEE International Conference on Systems, Man and Cybernetics, pp. 973–978].

4. Opportunity

The good news is that integrity limits can be anticipated by proofreading all the software code involved. Proof reading assesses the code as text similar to how an editor proofreads a manuscript regardless of the message in the manuscript. When used with test and evaluation it can reduce the 40% of time and money typically wasted by the T&E staff while waiting on fixes to problem reports then performing regression testing. Even then T&E finds only about 98% of the problems [www.namcook.com]. The other 2% are in the deployed system.

Whatever the cause most currently deployed software contains an unacceptable number of latent bugs. When users encounter these bugs during operations the maintenance staff can use proof reading to cut the problem diagnosis time to 10% of current experiences and preclude the typical case wherein approximately 5% of the fixes to one problem causes a collateral problem. Some testing is still necessary for discovering dynamic limits but can be accomplished in much less time and cost once the system is cleansed of integrity faults.

This is not about assessing just one program but about assessing the viability of hundreds to thousands of interoperating programs throughout a system that participate in producing an intended result or not. When logic, arithmetic and semantic faults can be identified as the source of observed error then the relevant thread through many modules can be highlighted for the edification of the software personnel responsible for devising a fix. Further, all this can be automated such that the task is performed in seconds rather than hours or days. Acceptance of proposed fixes can be done in seconds as well.

Beyond the software code it is likely that these techniques can be applied to assessing the integrity of the system model once it has been expressed in a formal, model-based language. This means that system design and architecture can be vetted from Day 1 of a project through Year N of its system usage life.

Beyond the system design it also means that we may be able to assess business processes and policies before these are embedded in the Enterprise IT system.

5. Software Integrity Assessment Value Proposition

The high incidence of faults in computer software is causing losses in the trillion dollar range throughout industry and governments worldwide. Symptoms include lost system effectiveness, lost data, lost time, and lost customer loyalty as well as vulnerability to human error and cyber actions. The situation inexorably worsens despite billion dollar expenditures on testing and "bug fixing."

Now these losses can be prevented because Software Integrity Assessment, SIA, can automatically detect all logic, arithmetic and semantic faults within a computer program and among associated programs ranging from applications to data base management to device handlers.

Preliminary estimates indicate that each dollar invested in the capability to assess integrity will return more than 5 dollars in reduced software costs and increase system operational availability --- a low risk, high ROI proposition. Also significant labor is freed up to pursue other system projects.

6. The Technology

Proofreading software code as text involves 3 steps. First, taking the software in whatever language it is expressed (the executable form is the least ambiguous) and converting it to Dijkstra Guarded Commands with the sequencing of state changing instructions removed (as introduced by K. Mani Chandi and J. Misra in UNITY). This is essential to allow assessment of liveness and composition of programs in a system such as creating a system of systems. Second, determining what the User wants to know. Typically expressed in terms of post conditions expected on subsets of system states, questions are;

A. This program produced an erroneous output. Where is the fault?
B. This program produced an error message. Why?
C. This program crashed. Why?

D. This program hung or went into a continuous loop. Why?
E. Has a change elsewhere violated an invariant for this result?
F. Are the existing invariants compositional?
G. Can this new set of code be added safely?
H. Which code is relevant for computing this result?
I. Which code is relevant for obtaining this data?
J. What does this program do?
K. Under which condition does this program do something?

Third, from the User's desired post-condition indicated by the question, calculate the valid Weakest Precondition, *wp*, else the specific fault.

One way of doing this is described in US Patent 6,029,002 dated 2/22/2000. A significantly different and quicker way of processing the code has evolved since 2008 with the advent of OntoPilot LLC.

7. Action

Software Integrity Assessment is key to achieving cyber-resilient systems. Most cyber remedies focus on intrusion prevention and information assurance. System Integrity Assurance, SIA, focuses on making the code fault-free thereby precluding the favorite modus operandi of cyber attacks which is to exploit these vulnerabilities. No vulnerabilities, no exploitation.

INCOSE members at OntoPilot LLC intend to help create the SIA talent base and tools. We intend to put the theory and method into the public domain, spawn several enterprises to commercialize SIA, encourage professional and industrial associations to add SIA to their certification programs and convince academia to add SIA to their curricula.

We estimate that approximately 100,000 SIA-adept practitioners will be needed to forestall deployments of latent bugs and to work off what is now called Technical Debt, the millions of latent bugs already deployed.

SIA know-how will be available to all interested parties and the respective parties will be encouraged to form a mutually leveraging ecosystem of automation development.

Fortunately, SIA is not a fork in the road. SIA can be easily integrated into most any software engineering practice or Critical Infrastructure Improvement practice thereby making the participants more productive, innovative and appreciated.

In addition to assessing all 16 Critical Infrastructures it is important that we also assess the 17th Critical Infrastructure, notably, the sociotechnical system we must organize and ensure as fault-free for assessing the other 16. Directives are necessary but not sufficient. It takes intelligent action.

All Critical Infrastructure that contains or relies on any programmable device will be made much less vulnerable. Let's do this.

<end>

Resilient Hospitals Handbook

Prepublication Release
For the Conference and Annual Meeting of the
Society for Disaster Medicine and Public Health
September 29, 2017
Newport News, VA

Resilient Hospitals Handbook

Strengthening Healthcare and Public Health Resilience in Advance of a Prolonged and Widespread Power Outage

By:

Charles (Chuck) Manto,

Earl Motzer, Ph.D

James Terbush, MD, MPH

Funding and support provided by the U. S. Department of Homeland Security, National Protection and Programs Directorate, Office of Infrastructure Protection, through the Technology Development & Deployment Program (TDDP) which is implemented by The National Institute For Hometown Security (NIHS).

Funding and support provided by the U. S. Department of Homeland Security, National Protection and Programs Directorate, Office of Infrastructure Protection, through the Technology Development & Deployment Program (TDDP) which is implemented by The National Institute For Hometown Security (NIHS). A published edition of this will be available through Westphalia Press and Amazon by the Fall of 2017.

Resilient Hospitals Handbook: Strengthening Healthcare and Public Health Resilience in Advance of a Prolonged and Widespread Power Outage[1]

Authors

Charles (Chuck) Manto, President Instant Access Networks, Chair EMP Special Interest Group InfraGard, Co-principal Investigator (cmanto@stop-EMP.com);

Earl Motzer, Ph.D., Chair, Healthcare and Public Health Sector Coordinating Council, Principal Investigator;

James Terbush, MD, MPH, Public Health Instructor and President, Innovative Health Systems Inc., Co-principal Investigator

Abstract

A number of high-impact threats to critical infrastructure can result in a regional or nationwide months-long power outage, making it unlikely for timely outside help to arrive. Hospitals are encouraged to gain the capacity to make and store enough power on-site to operate in island mode indefinitely without outside sources of power or fuel and protect on-site capabilities from threats that could impact regional commercial power systems. This handbook outlines challenges and opportunities to solve these problems so hospitals, healthcare facilities, and other resources might become more resilient.

Introduction

In August 2005, Hurricane Katrina struck New Orleans and subsequently became one of the most destructive and expensive disasters in history. Hurricane Harvey's recovery costs in 2017 and beyond could be larger. Fortunately, disasters of this magnitude are rare, and some of the worst consequences may be preventable. Although loss of life has been statistically decreasing in domestic disasters (fortunately), the cost of disasters is skyrocketing. Some types of disasters, however, can still cause catastrophic loss of life. Disasters of the type discussed in this handbook may occur only once in a generation or even once in a lifetime.

---

[1] The unrestricted general release of this report is in partial fulfillment of a Resilience Challenge in partnership with the National Institute for Hometown Security funded by the U.S. Department of Homeland Security.

Likelihood. The chance of just one type of these disasters occurring, extreme space weather in particular, has been described by various experts as having a likelihood of 6%–12% in any decade.[2] A small number of high-impact, low-frequency (HILF) threats can deliver similar effects on their own. Any one of these threats or combination of threats may result in a widespread long-term collapse of critical infrastructure, which could minimize the ability of outside help to the affected communities. Although society has no way to stop natural hazards such as solar storms from occurring, there is much hospitals can do to minimize the consequences to their facilities. On the other hand, the chances of intentional man-made events decrease to the extent protections are in place and increase to the extent they are not in place.

Impact. According to various reports, including Lloyds of London,[3] a 100-year storm could result in power outages lasting 1–2 years from DC to NYC alone.[4] Described as HILF events, they can even be existential in nature: for a particular business, for a specific hospital, for a community, or for a region. In turn, the severity of these threat scenarios compels local communities and institutions to be more resilient, so they can perform through these events and be in position to help their neighbors.[5]

Significance of healthcare. Of the 16 Critical Infrastructure Sectors[6] (as described in the National Response Framework), Healthcare and Public Health is perhaps the most fragile sector and is the most dependent on all the other sectors. Healthcare as practiced in the United States is complex and becoming more so as healthcare facilities become ever more dependent on technology, especially with regard to the U.S. power grid and information technology (IT). Hospitals cannot function without a steady uninterrupted supply of power and access to the digital world.[7] This handbook familiarizes and orients the reader to the subject of a prolonged and widespread power outage affecting all infrastructure sectors. However, the focus here is on threats to Healthcare and Public Health critical infrastructure, specifically hospitals and other large, fixed-base healthcare-related institutions. The purpose of the handbook is to help educate healthcare facility owners and operators (especially chief executive officers and their boards of directors) and other leaders on how to maintain resilience of their institutions throughout a high-impact event, specifically a prolonged and widespread power outage, in a cost-effective manner.

This handbook is *not* particularly focused on cybercrime, hacking, or cyberterrorism except as it relates to threats against the power grid. It will, however, address other scenarios that are typically beyond the scope of The Joint Commission Organization[8] review or a typical hazard vulnerability analysis. It is also

---

[2] See Pete Riley's article in Space Weather and abstract at: http://onlinelibrary.wiley.com/doi/10.1029/2011SW000734/abstract "the probability of another Carrington event (based on *Dst* < −850 nT) occurring within the next decade is ~12%."

[3] See the Lloyd's of London and AER 2013 report, "Solar Storm Risk to the North American Power Grid": https://www.google.com/search?q=solar+storm+Lloyds+of+London&oq=solar+storm+Lloyds+of+London&gs_l=psy-ab.3...613573.622915.0.623282.30.28.1.0.0.0.223.3091.3j19j2.24.0....0...1.1.64.psy-ab..12.17.2090...0j35i39k1j0i20k1j0i131k1j0i3k1j0i22i30k1j0i22i10i30k1j33i22i29i30k1.IoPK__ApmAo

[4] See impacts up to 1–2 years on coastal areas including the Gulf Coast and the Midwest.

[5] These same actions can also create a more secure local economy and environment.

[6] https://www.dhs.gov/critical-infrastructure-sectors

[7] Water is also critical and deserves special attention as illustrated by Baptist Hospital in Beaumont, Texas, closed by lack of water during Hurricane Harvey. Water utilities are similarly dependent on power. Hospitals need to work with their water utilities to develop greater resilience and develop on-site alternative sources whenever possible, enhancing the benefits of microgrids using geothermal systems and pumped storage facilities as energy storage. In many cases, power may be the infrastructure with the longest lead time to repair and, hence, the more urgent to develop and protect from regional grid threats.

[8] www.jointcommission.org/

not intended to be an exhaustive source of references on the subject of hospitals and disasters, nor will this handbook specifically address all requirements of the new Centers for Medicare & Medicaid Services' (CMS) Emergency Preparedness Rule.[9]

The handbook does, however, function as a checklist with a set of questions for the reader to address well before a disaster. The primary question is how to create, store, and use the power necessary to maintain essential functions and life safety for the hospital facility, which may be required to thrive and provide a surge of additional services in such a disaster. The five scenarios addressed here (i.e., cyberattack on the grid, extreme space weather, pandemic affecting a large percentage of the U.S. population, physical attack on the power grid, and electromagnetic pulse [EMP]) are unfamiliar topics for many hospital emergency managers. The common thread is how each of these scenarios affects the healthcare system, either directly or indirectly.

The handbook's content is specifically meant to be outside the usual set of challenges posed to hospitals in preparing for certain disasters and to stimulate new ideas and possible solutions. Hospital leaders can take some practical measures now to reduce loss of life and limit risk to their institutions in the event of a prolonged and widespread power outage, which are detailed later in this report. Preparedness efforts before a disaster facilitate a superior and much more cost-effective response than a response without adequate preparedness. Many of the solutions for high-impact threats can provide day-to-day benefits and resilience in the face of other unforeseen challenges.

Brief Overview of the U.S. Power Grid: How It Works and What It Looks Like

The North American power grid has been described as the largest and most complex machine in the world. Consisting of eight regional entities, it covers the bulk power supply of both Canada and the United States. Within those regions are over 1,400 independent and semi-independent power companies under the oversight of two independent regulatory bodies—the Federal Energy Regulatory Commission (FERC) and the Nuclear Regulatory Commission (NRC)—as well as the Public Utility Commissions in each of the 50 states. To ensure coordination among the many thousands of power providers, the North American Electric Reliability Corporation (NERC) develops and enforces grid reliability standards. Power comes from a variety of sources: coal- and natural gas-fired utilities, nuclear, hydroelectric, solar, wind, and others. These power sources are connected by hundreds of thousands of miles of high-voltage and lower voltage electric lines. Balancing the loads and ensuring each jurisdiction is supplied according to the demand are functions of 16 NERC balancing authorities. Finally, the eight regional entities are grouped into three major regions in the United States and Canada: the Eastern Interconnection encompasses the area east of the Rocky Mountains and a portion of the Texas panhandle; the Western Interconnection encompasses the area from the Rockies west; and the Electric Reliability Council of Texas (ERCOT) covers most of Texas.

Transformers at the various substations first step up the voltage for efficient transmission of power across long distances and then again step down the voltage for use by customers. Most of the extremely large high-voltage transformers are custom-made, weigh hundreds of thousands of pounds, are largely sourced from overseas, and are unique for their purposes. At various points in the distribution of electricity are monitoring and industrial control functions, some at remote locations. These are frequently unprotected industrial controls or automated Supervisory Control and Data Acquisition (SCADA) systems. These systems are often legacy devices that are accessible through the Internet, embedded

[9] www.cms.gov/Medicare/Provider-Enrollment-and-Certification/SurveyCertEmergPrep

throughout the grid, and may not have been designed with common computer security measures in place. SCADA vulnerabilities are well known to the industry and make it possible for "hackers" to access essential grid functions.

Electromagnetic vulnerability. Intense solar activity (solar storms) has been shown to cause a reaction of the Earth's electromagnetic field electrifying the ground delivering electricity to travel along pipelines, rails, ground lines, and above-ground transmission lines; and inducing currents large enough to damage or destroy grid components (e.g., transformers). Although these storms may take hours to days to reach Earth, "turning off" the grid may not be possible or sufficient to avoid damage. Some types of readily available means of protection could be put in place to protect much of the grid against solar storms, but, for the most part, have not occurred.[10] Other physical phenomena associated with either nuclear EMP or non-nuclear electromagnetic weapons create frequency speeds in billionths and trillionths of seconds, faster than lightning arrestors function. These electromagnetic forces, arriving in nanoseconds, are much more destructive to electronics and require more sophisticated means of protection.[11]

Physical vulnerability. Almost all of the large high-voltage substations and transmission lines are above ground. Some are visible to passersby and are often behind a chain-link fence. As such, they are vulnerable to physical attacks such as happened recently in California.[12] Finally, people ultimately control the power grid. Anything that prohibits employees to get to their places of work can affect the power grid.

Examples of Prolonged Loss of Grid Power to U.S. Hospitals

Hurricane Katrina, August 2005. What happened at Memorial Hospital in New Orleans, Louisiana, after Hurricane Katrina is instructive for what could happen elsewhere in the event of a prolonged power outage. One book, entitled "Five Days at Memorial: Life and Death in a Storm Ravaged Hospital,"[13] provides an emotionally engaging and in-depth picture of what it was like without electricity, running water, air conditioning, or supplies. To quote now from a *New York Times* August 2009 article[14] by Dr. Sheri Fink,[15] "As at many American hospitals in flood zones, Memorial's main emergency-power transfer switches were located only a few feet above ground level, leaving the electrical system vulnerable." "'It won't take much water in height to disable the majority of the medical center,' said Memorial's CEO." What is interesting is that, unlike some other disasters affecting other hospital's building structure, Memorial was left mostly intact by the hurricane. Loss of emergency power was the critical feature. The Joint Commission requirement is for 96 hours of emergency power backup, but this

---

[10] At the time of this handbook's preparation, various power grid elements continue to be tested by government and industry programs. Some utilities have begun to take efforts to protect portions of their infrastructure. However, the warnings of the Defense Threat Reduction Agency (DTRA) and others still stand.

[11] The military has been protecting some of its systems since the 1950s.

[12] One of the most well-known recent attacks was the one on the Metcalfe substation in April 2013 (see https://en.wikipedia.org/wiki/Metcalf_sniper_attack).

[13] See https://www.amazon.com/s/ref=nb_sb_ss_i_2_14?url=search-alias%3Dstripbooks&field-keywords=five+days+at+memorial+life+and+death+in+a+storm-ravaged+hospital&sprefix=Five+days+at+m%2Caps%2C224&crid=3OAEB5QWYKMFG

[14] See August 25, 2009 NYT Article, "The Deadly Choices at Memorial": http://www.nytimes.com/2009/08/30/magazine/30doctors.html?pagewanted=all&mcubz=1

[15] See this February 13, 2016 NYT article challenging hospital preparedness in general, "Can Health Care Providers Afford to be Ready for Disaster?": https://www.nytimes.com/2016/02/14/sunday-review/can-health-care-providers-afford-to-be-ready-for-disaster.html?mcubz=1

typically only supports emergency lights, some equipment such as ventilators, and a few designated outlets throughout the hospital. The air conditioning system at Memorial shut down as did the elevators in the last few days of August. Supplies, food, and pharmaceuticals were all in limited supply afterward as the hospital was cut off by the floodwaters. After approximately 48 hours, the backup generator failed due to rising water and, sometime after that, potable water to the hospital also ceased. By September 1, "Dr. Pou was informed by the Tenet incident commander that 'no help was coming' and apparently the hospital was to be abandoned."[16]

As detailed in Sheri Fink's account, some patients lingered in these horrific conditions and ultimately succumbed to the heat.

The difference between Memorial Hospital and hospitals affected by Hurricane Katrina is that there was still the possibility of evacuation. Most of the patients, in fact, were successfully evacuated. Much of the effort directed toward saving patients was to transport them quickly to another facility by boat or helicopter. Evacuation of all patients was seen as the last best resort for a hospital that was failing. However, what if evacuation could not be accomplished due to lack of transport or lack of other functioning facilities within a reasonable radius? In a widespread and prolonged power outage, there would not be an evacuation option. Other medical facilities in the affected region would also plan to evacuate, but to where?

Super Storm Sandy, October 2012.[17] During Super Storm Sandy, more than 8.5 million households and businesses in the Northeast United States were without power from the grid. Hospitals in New York City had gone to great lengths before the storm to update and improve emergency power generation. For the New York University (NYU) Langone Medical Center,[18] it was not enough. When sensors in the fuel tanks supplying those generators detected water from the storm, they shut down. Without emergency power, NYU Langone was forced to evacuate hundreds of patients to nearby hospitals. Emergency generators of all types, throughout the region, required refueling after several days and required maintenance to keep them running. Frequently, the same fuel suppliers were contracted to provide emergency fuel to many locations and were unable to keep pace with demand. When roads were closed and fuel was at a critical level, Department of Defense (DoD) assets were requested to provide relief.[19] Assets from as far away as California assisted, something far less likely during a nationwide prolonged power outage.

Although widespread, the power outage was not particularly prolonged for most customers. In this case, area hospitals were again called on to provide bed space for NYU Langone patients. Long lines of ambulances waited to take patients to nearby hospitals that still had power. The evacuation was deemed successful.

---

[16] See the rebuttal of certain items in Dr. Fink's book by Dr. Pou at this website: http://www.memorialhospitaltruth.com/whathappened.html

[17] Sandy quick facts: Impacts Jamaica on October 19, 2012, and Atlantic City, NJ, October 29, 8.5 millions without power on October 31, half restored by November 1 (see Live Science at https://www.livescience.com/24380-hurricane-sandy-status-data.html).

[18] Since Hurricane Sandy, NYU Langone has developed considerably more resilient local power infrastructure, including a new combined heat and power system powered by natural gas. However, that system and the gas company supplying the natural gas are not EMP-protected and remain vulnerable to many of the same threats facing regional grids.

[19] Fortunately for many, the U.S. DoD was able to supply necessary resources from around the country including the West Coast.

In a truly widespread and prolonged power outage, however, evacuation would not be an option. In addition, most hospitals already operate at or near capacity. Finding open beds, particularly for critical or trauma patients, is an almost daily struggle for some facilities. Worst-case scenarios for various disasters call for medical support for an additional 200,000 or more citizens in a major metropolitan area, thus requiring substantial surge capabilities and support from functional facilities.

Countering expectation. When threats impact the bulk of the country simultaneously, Federal Emergency Management Agency (FEMA), Department of Health and Human Services (HHS), Department of Homeland Security, and DoD capabilities are likely to be stretched thin and may not be available for a considerable time, if at all. In such situations, hospitals need to shelter in place or cease providing medical care altogether when a situation becomes untenable. Of course, it would be far better if every institution were so resilient they could continue to maintain close-to-normal capabilities and even meet the surge of additional demand that is likely in those circumstances.

Possible HILF Threats Affecting the Power Grid

Cyberattack on the U.S. power grid. This has already occurred and is very likely to occur again. In 2014, Russia tested a complex attack scenario (HAVEX/Black Energy2) against the American Grid. The attack primarily affected Industrial Control Systems (ICS) and SCADA devices. Subsequently, Russia was able to launch its improved version of the malware on the Ukrainian national grid, shutting down power to about 230,000 customers. The attack included a distributed denial of service attack on call centers. The attack did not damage expensive and difficult-to-replace equipment, but rather demonstrated the ability to take complete control over the energy sector control centers. Much earlier, in 2007, Idaho National Laboratories demonstrated the ability of a cyberattack to cause physical damage (Aurora Test) to a diesel generator by opening and closing circuit breakers remotely. Since that time, cyber weapons have become increasingly sophisticated and countermeasures continue to be implemented but appear to lag behind the threats. The Secretary of Energy, under the FAST Act of 2015, cautions against minimizing the ongoing vulnerabilities of the U.S. energy sector to cyberattack.

Some steps to protect against cyberattack on the grid include concepts that could be applied to local power generation sources such as microgrids:

- Encrypt all control center communications
- Limit on-site access to data systems
- Require biometric means for access to certain areas
- Use two-factor verification for all systems
- Limit remote access
- Reduce equipment having a single point of failure
- Minimize Internet access to SCADA and control systems
- Enhance cybersecurity on all systems
- Provide additional cybersecurity training to all employees
- Establish 24/7 cybersecurity monitoring
- Use only "whitelisted" equipment for SCADA and ICS
- Do not rely on "air-gapped systems" to remain offline
-

Solar storms or extreme space weather. This is a fascinating area that continues to motivate scientific investigation with substantial real-world consequences for the power grid. In 1989, a relatively minor solar event interrupted power for over 6 million customers in Canada for at least 9 hours. In 1859, the so-called Carrington Event struck North America with a walloping solar storm and induced a geomagnetic disturbance (GMD), which damaged telegraph lines and some telegraph stations across the continent. The sun randomly continues to produce coronal mass ejections (CMEs) sufficient to physically destroy much of the power grid the nation relies upon. The only defense so far has been that the Earth was not in the direct path that the directional CMEs could cause much damage.

What many may not realize is that, on July 23, 2012, the Earth narrowly missed a major coronal mass ejection from the sun by about 9 days.[20] This was subsequently described as a Carrington-level storm that, according to estimates by Lloyds of London, could create a power outage lasting as long as 18 months from NYC to DC.[21]

Imagine if Earth had been hit by that massive solar storm in July 2012, only to be hit by Hurricane Sandy the following September, long before being able to recover from the larger power outage caused by the giant solar storm. Relatively little help would have been available outside the region affected by Sandy since many areas outside the hurricane-impacted area would have still been coping with their own difficulties.

Long-term "grid" vulnerabilities for solar storms include the destruction items requiring long lead times, including large transformers that transfer power across continental power distribution regions. Since transformers take over a year to build and procure, with few available spares, their destruction could result in extended lead times for custom replacements to be manufactured and reinstalled.

Hospital executives are usually not in the position to take measures to harden the bulk American power system, but they are in the position to make enough of their own power to maintain hospital operations. In fact, even when the regional power grids are protected from various threats, being overly dependent on centralized infrastructure is like putting all the eggs in a few giant technology baskets. It makes more sense for many reasons to diversify risk and increase resilience by making critical loads locally on or near a particular location.[22]

Electromagnetic pulse. EMPs result from a number of causes, the most impactful being a nuclear detonation in the upper atmosphere. High-altitude weapons testing in the 1950s and 1960s demonstrated that the phenomenon could cause significant damage to electronic equipment, but without ionizing radiation, blast, or heat on the ground. Older technology (i.e., vacuum tube) was more resistant to the extremely powerful and extremely fast EMP than the microcircuits contained in virtually all modern devices. A single explosion (or several explosions) high above the United States could cover the country coast-to-coast. The commercial power grid is not yet EMP protected and remains a key vulnerability that could cause cascading effects to other systems, including communications, water, and healthcare facilities.

---

[20] See https://en.wikipedia.org/wiki/Solar_storm_of_2012

[21] See https://www.google.com/search?q=solar+storm+Lloyds+of+London&oq=solar+storm+Lloyds+of+London&gs_l=psy-ab.3...613573.622915.0.623282.30.28.1.0.0.0.223.3091.3j19j2.24.0....0...1.1.64.psy-ab..12.17.2090...0j35i39k1j0i20k1j0i131k1j0i3k1j0i22i30k1j0i22i10i30k1j33i22i29i30k1.IoPK__ApmAo

[22] Some measures which could protect the power grid against extreme solar weather are:

- Install GMD monitoring and blocking devices at various select critical locations
- Provide additional spare equipment to selected substations
- Standardize major equipment to facilitate replacements
- Stockpile replacement transformers in select locations
- Shed load, if and when possible, to reduce transformer reactive power consumption and overheating

Coordinated physical attack on the power grid. This continues to be a significant threat. Although great strides have been made to ensure the electric grid's reliability, so too have the nation's adversaries who have sought to exploit inherent vulnerabilities. Although EMP, solar weather, and cyberattacks against the electric grid can be somewhat complex, a simple physical attack on a substation can have a similar impact. The vulnerable points to an electric grid physical attack include power generation plants, transmission lines, transformer substations, utility-owned communications systems, and industrial control facilities.

Most power generation plants are physically manned with defined defense perimeters such as chain-link fences and video camera monitoring. Notwithstanding security measures, successful physical attacks occur. One case in point is a sniper attack on the Metcalf Transmission Substation located in Coyote, California, near the border of San Jose. It occurred on April 16, 2013, when gunmen fired on 17 electrical transformers, which affected the greater San Jose area.[23]

Before the shooting, the attackers cut a series of fiber-optic telecommunication cables. It was a sophisticated and well-planned attack with the apparent intent of creating an extended electrical power outage in one of the nation's high-technology centers. The extended power outage, albeit targeted toward the area industrial complex, would have affected all electricity consumers, including hospitals. During the incident, a Pacific Gas and Electric (PG&E) technician was coincidentally performing testing on the electric grid and noticed an abnormality coming from the Metcalf substation. Had he not immediately implemented emergency procedures and had alternative routing capability, an extended and widespread electrical power outage, due to cascading failures of adjacent power stations, would possibly have occurred.

The incident was an eye-opening event for PG&E, which subsequently invested over $100 million to protect the substation against unauthorized entry. Unfortunately, the Metcalf substation was again breached in August 2014 and further damage to substation components was inflicted. To date, both breaches remain unsolved.

The Metcalf substation event illustrates the fact that events can and do happen outside an organization's control. The lesson to be gained is how best to prepared and build resilience to ensure organizational survivability and continuity of operations.

Pandemic (highly pathogenic influenza Type A). This seems to be the outlier in this HILF list. In fact, some experts believe that a severe pandemic is a near certainty, "not if but when," in the next few decades. Extrapolating from the 1918 Influenza pandemic to today, approximately 90 million Americans would become ill (assuming a 30% attack rate), almost 10 million would need to be hospitalized, 1.5 million would need care at an intensive care unit, and 750,000 would need mechanical ventilation. Clearly, the capability is not currently available to care for such numbers. More to the point, it can be assumed that perhaps 50% or more of those who become ill would seek medical care. This would overwhelm existing capabilities, hence a recommendation for many persons with influenza-like illness to stay home unless the patient meets certain criteria.

DoD used a working estimate of 40% absenteeism from work during a pandemic[24] as the threshold for critical infrastructure to begin to fail. Many employees would remain home to care for a loved one or to avoid getting the infection from co-workers. Employees of the energy sector (power grid) are no

---

[23] See this 2016 article reviewing Metcalfe and concerns about similar attacks: https://www.eenews.net/stories/1060043920

[24] This includes work led by Dr. Terbush.

exception. When absenteeism hits 40%, it is estimated that sections of the power grid would go offline. When that happens, the scenarios of previous HILF threats and consequences for hospitals become the same or similar.

Some measures available to prevent and/or limit the spread of pandemic Influenza include:

- Routine use of hand-washing, cough hygiene, and social distancing measures
- Support for new rapid vaccine development technology
- Widespread distribution of (effective) antiviral medications
- Consideration of school closures (children are more efficient spreaders of Influenza A)
- Limitations on large gatherings (e.g., conferences, shopping malls, and church services)
- Telework (if available, telework has been shown to significantly decrease the spread of illness)
- Home care (stay at home if ill and avoid going to the hospital unless certain criteria are met)
- Prevention resources (e.g., CDC website, https://www.cdc.gov/flu/pandemic-resources/index.htm)
-

Other weapons of mass destruction. These can also function as high-impact threats to critical infrastructure but are not examined in this handbook. However, resilience that provides protection from high-impact threats covered in this handbook would provide benefits for other threats and scenarios that are not specifically addressed.

Recommendations for Protection Against All High-Impact Hazards Threats to Power Infrastructure

Although eliminating the risk of attack on the electric grid is impossible and outside the responsibility of any individual organization, it is possible to mitigate the indirect effects if (when) an attack against the electric grid were to occur. Hospital executives can enhance protection for a local hospital facility against attack on the electric grid by a variety of means such as:

Current Power Systems

- *Procure redundant power via electrical feeders from different substations and, if possible, different providers.* Receiving electrical power from different providers diversifies electrical power sources and thereby mitigates some risk.
- *Test generators under full load.* Many organizations test their generators on a monthly basis to determine if they will start and operate as desired. If diesel generators are repeatedly started but not under a full load, full output may not be realized when needed. Generators available to hospitals can include monitoring and testing systems to ensure they work when needed. In long-term power outages, diesel generators would still be useful but insufficient to guarantee power availability.
- *Enhance backup power generation.* Diesel generators are the most available and a proven source of backup power that should be maintained and exercised. However, they have severe limitations in an extended power outage scenario. Diesel generators must have sufficient backup power to meet critical mission areas. Equally important, facilities must have sufficient diesel fuel supply to endure for an extended power outage. Hospitals are required to have 96 hours of fuel, but 7 days is a more realistic minimum standard for smaller day-to-day outages. If an extended power outage occurs, contracted fuel delivery would be assigned to the highest priority requirements and redirected by local and federal officials. In extended power outages, refineries may cease functioning and little fuel may be available at all. Fortunately, the infrastructure of backup power can be used to form the design basis of an on-site or near-site microgrid.

New Systems

- *Create an on-site (or near-site) power generation and storage capability incorporating on-site energy sources with an appropriate level of energy storage to provide the base loads required to keep the hospital functional (usually 30%–50% of all power).* Renewable energy such as wind, solar, and geothermal systems can diversify electrical energy sources and enhance resilience. The challenge is to have sufficient energy storage such as batteries to meet critical needs at all times including in island-mode detached from the grid. This solution (i.e., renewables and energy storage) can serve as an uninterruptable power supply (UPS) for the base load that can continue to operate through the power outage, thus allowing backup diesel generators to either be kept in reserve to support the microgrid or provide power over and above what the microgrids would provide in the case of a prolonged outage.
- *Consider energy savings from local power generation and energy efficiency improvements to help discover funding for new microgrids.* The energy produced by local microgrids can produce energy savings that can be used in power purchasing agreements that can help pay for the procurement of EMP-protected microgrids. Coordination with other energy efficient systems can further fund microgrids out of savings.

- *Consider natural gas-fueled generators.* Natural gas generators can provide a suitable alternative electrical power energy source. However, they may be vulnerable to a widespread electrical power outage, where natural gas pipeline pumps are powered from the electric grid. However, natural gas generators (usually in configurations as a combined heat and power [CHP] system) can be part of an overall power solution that can help fund the development of a more complete system, a portion of which can be more resilient than the unprotected CHP system.
- *Consider creation of EMP-protected facilities for power, network, and data center infrastructure.* The development of a control room for a new microgrid could be expanded to provide protection for other networking, data center, and security needs.
- *Consider developing a system-of-systems evaluation of the facility.* This would be similar to those supported by the NIHS program producing this handbook, which could provide technical and financial support necessary to enhance resilience in light of high-impact threats.
- *Consider inclusion of systems modeling and management.* This could aid in the development of a microgrid, help manage a microgrid after it is built, and assist in its ongoing development, including the potential to sell power in coordination with local utilities and support connecting and disconnecting from power grids.

All Systems

- *Bolster physical security.* Electrical feeder lines and backup power generation are often secondary considerations when considering physical security even though they represent key vulnerabilities to operations. Security cameras and motion detectors should also be considered to bolster physical security measures and be part of an independent EMP-protected power and network management system that continues to function throughout a power and network outage.
- *Limit access to essential personnel.* As electrical feeders and backup power generation are critical nodes to operations, access should be limited to only trusted personnel who have a requirement to maintain and/or test facilities. All personnel with access to critical facilities (including custodial and contractors) should be screened with at least minimal background checks to minimize the potential for "insider threats."

Why Focus on EMP?

Focusing on an EMP is useful for a number of reasons. First, it can occur in more ways than most people are aware, and some EMP events can create the most widespread and long-term impacts to critical infrastructure such as power and water. Second, there are proven ways to address EMP through island-mode operation, which provides a helpful resilience model for facility owners and hospital managers.

It helps to understand that disruptive or damaging electromagnetic interference (EMI) can occur in a number of ways. Precautions against the harshest intentional manmade versions provide protection against the others. Nature provides its own versions from extreme space weather. Manmade EMI can be accidental or intentional.

Intentional EMI can come from small electronic weapons that can be carried in briefcases or larger ones that can be driven by van or truck. Now that capabilities of the weapons are increasing, it is possible to drive by a targeted control or data center and disable it without ever getting closer than a few hundred yards outside a security fence. Perpetrators could deliver damaging attacks without being noticed and continue to the next target (or come back another day). Motives for that kind of attack range from terrorism to sophisticated criminal actions (e.g., to short the stock of a targeted public company).

The most impactful EMP results from the detonation of a nuclear weapon 10–300 miles aboveground, which then creates electromagnetic fields that stream to Earth in all directions.[25] The higher the detonation, the larger the impacted area would be. Because there is no need for vehicle reentry into the lower altitudes or great requirement for accuracy, experts believe that this could be an easier technology to develop or acquire for emerging nuclear powers as well as non-state actors. (See EMP SIG bibliography.) A small weapon detonated 80 miles high could affect a large multistate region, whereas one detonated 300 miles over the center of the United States could affect the entire continental United States. A ground-burst nuclear weapon would have a more limited radiated EMP, but an EMP that travels along conductors such as power lines could be more far-reaching and affect a larger area.

Both non-nuclear and nuclear EMP could disrupt or damage power generating stations, SCADA devices, computer equipment, and telecommunications networks. The proven ways to protect facilities from EMP include resilience of local facilities to operate in island-mode through the initial attack and until recovery is complete.

The likelihood and severity of either a nuclear or non-nuclear EMP attack was sufficiently large for the Defense Threat Reduction Agency (DTRA) to recently launch a request for proposals through the Small Business Administration program named the Small Business Innovation Research Program. The proposal seeks to create EMP-protected microgrids. In its official announcement and subsequent press release approved for unlimited general release, it stated, "An electromagnetic (EM) attack ... has the potential to degrade or shut down portions of the electric power grid important to the DoD.... Restoring the commercial grid from the still functioning regions may not be possible" (See the RFP from DTRA).[26] The DTRA goal was to "accelerate the adoption of EMP-protected critical infrastructure and microgrids among civilian institutions that need to operate in island mode during a prolonged power outage. These include... water utilities, hospitals and emergency communications."

A further risk is damage or destruction of micro-electronic circuitry in any of the various devices that people have come to heavily depend on for daily activities, especially larger items and those connected to wires of a meter or longer.[27] Medical devices in hospitals are not exempt from this effect. In addition

[25] Even a small 5–10 kiloton weapon can create a significant EMP depending on emissions.

[26] See text within RFP: https://www.sbir.gov/sbirsearch/detail/736859. "An electromagnetic (EM) attack (nuclear electromagnetic pulse [EMP] or non-nuclear EMP [e.g., high-power microwave, HPM]) has the potential to degrade or shut down portions of the electric power grid important to the DoD. While a power grid may employ intentional islanding techniques to protect sections of the grid and prevent a cascading collapse of the power grid, the broad reach of potential EM attacks with the potential of simultaneous levels of disruption might prevent traditional islanding protection methods from being sufficient for continued operations of the DCI. Restoring the commercial grid from the still functioning regions may not be possible or could take weeks or months. Significant elements of the DCI require uninterrupted power for prolonged periods to perform time-critical missions (e.g., sites hardened to MIL-STD-188-125-1). To ensure these continued operations, DCI sites must be able to function as a microgrid that can operate in both grid-connected and intentional island-mode (grid-isolated). Such a microgrid is defined as a group of interconnected loads and distributed energy resources within clearly defined electrical boundaries that acts as a single controllable entity with respect to the power grid."

[27] Another considered risk is that modern automobiles contain many microchips. In an EMP event, these microchips may stop the engine from running or even fail to start at all, especially if connected to a wiring system that would make it possible for the electromagnetic currents to couple with them. Changes in the use of automobile

to computers and communication equipment, IV pumps, bedside monitoring equipment, and ventilators could be affected as all contain microchips. Again, the size, configuration, and connections at the time of a pulse are all factors.

In a dystopian novel based on a severe EMP scenario (i.e., "One Second After," by William R. Forstchen),[28] three public health crises occur as a result. The first is lack of ability to access medical care (i.e., hospitals and clinics) and lifesaving medicines (i.e., antibiotics and insulin). Persons who rely on electricity-powered devices to keep them alive (i.e., oxygen generators and dialysis machines) succumb without access to power. The second is a period of civil unrest, which is outside the scope of this discussion. The third is a return to preindustrial methods of food production, food distribution, and economic exchange. All three are associated with significant hardship and loss of life. In a worst-case scenario EMP, a 1-year power outage (or even longer) would possibly occur before the majority of the power distribution network is restored.

Characteristics of Resilient Hospitals

Beginning with Homeland Security Presidential Directive 21 (HSPD-21) and followed by Presidential Preparedness Directive 21 (PPD-21), federal level planners have increasingly used the term "resilience." Resilience as an operational concept, however, has been around a long time. As a point of reference, "continuity of operations" can be used almost interchangeably. For the purpose of this handbook, a shortened definition of resilience is "the ability to take a blow and come back." The Joint Commission (JCO) established a baseline for emergency preparedness for hospitals and other healthcare institutions.[29] Although the specific requirements for the Healthcare Organizations' evaluation are available for purchase from The Joint Commission, there is a wealth of detailed helpful information on the JCO website for hospitals to become more disaster resilient.[30]

The CMS/HHS Final Rule for healthcare organizations also establishes national emergency preparedness requirements for Medicare- and Medicaid-participating providers and suppliers.[31] These requirements help providers and suppliers plan adequately for both natural and man-made disasters, and coordinate with federal, state, tribal, regional, and local emergency preparedness systems. Requirements are set for emergency power generation at many types of healthcare facilities within the Final Rule.

The HHS Sustainable and Climate Resilient Healthcare Facility Initiative is useful not only for weather emergencies but also for disasters of many types.[32] This online document and toolkit provides a broad look at healthcare facility resilience. Although primarily focused on disasters related to climate change and severe weather, the steps recommended for healthcare facilities to take to improve resilience would be appropriate and useful for any of the five HILF scenarios. The toolkit is divided into various elements

---

electronics, configuration, wiring methods, and auto body materials would all impact the likelihood and amount of impact for any particular vehicle. (*Note:* DoD funded the development of EM devices mountable in helicopters to disable automobiles driven on highways.)

[28] See the series of books by Dr. Forstchen at https://www.amazon.com/s/ref=nb_sb_noss_2?url=search-alias%3Dstripbooks&field-keywords=One+Second+After&rh=n%3A283155%2Ck%3AOne+Second+After

[29] Joint Commission (JCO) requirements can be found at: https://www.jointcommission.org/emergency_management.aspx

[30] https://www.jointcommission.org/assets/1/6/US_National_Library_of_Medicine_-_Disaster_Health_Information_Resources_052014.pdf

[31] https://www.federalregister.gov/d/2016-21404

[32] https://toolkit.climate.gov/topics/human-health/building-climate-resilience-health-sector

with Element #3 being Infrastructure Protection and Resilience Planning. Element #3 is further divided into the following "Steps": energy and utility infrastructure, energy conservation, water supply, water usage, sewage and wastewater, communications infrastructure, and medical information infrastructure. The document also introduces the term "islanding."

Resilient community islands. The concept originally began with the energy sector and the necessity of being able to "black start" some power stations (i.e., restart power stations after a widespread power outage). This, in turn, would help to bring larger sections of the power grid back online. The term "resilient community islands" has since been used to refer to resilient organizations: university campuses, state and local governments, military bases, and certain critical businesses, which each continue their essential functions and remain in communication with other "islands." Another example of a functionally differentiated community island would be a medical campus/hospital. If the hospital had its own power supply that could work without a functioning commercial grid and other essential resources (e.g., water, food, and communications), it could be a focus for medical care for other functioning islands and possibly help to bring other medical facilities back online.

Graceful degradation. The concept of "graceful degradation" of services is not new. Continuity of operations plans have long been developed to maintain essential services and to eliminate nonessential ones. This concept includes the idea that, in a widespread and long-term power outage, hospitals may need to continue to provide the most essential services for as long as possible and then have a plan for which services must be eliminated. Resilient hospitals have plans in place to fail gradually and not catastrophically (fragile systems). This requires thoughtful analysis of a particular facility's mission as well as the particular needs of the community the hospital serves.

Electricity is, of course, not the only critical resource needed to maintain hospital operations. Other critical infrastructures affecting hospitals include water and wastewater and sewage removal. In fact, water may be even more essential to hospital operations than electricity. A hospital would lose the ability to provide essential functions almost immediately if the water supply is interrupted when the grid goes down. Some specific requirements, such as loss of clean laundry, also have a disproportionate negative effect on hospital functioning. Other critical infrastructure sectors affecting healthcare include: communications, emergency services, transportation, and manufacturing.[33] A case can be made that all of the other critical infrastructure sectors affect the Healthcare and Public Health sector.

The most critical resource for maintaining hospital operations is the human resource. A 2015 study was conducted using a variety of different disaster scenarios that asked the question, "Will hospital staff come to work?"[34] The results were perhaps surprising and disappointing. The most commonly cited reason for not coming to work was concern for family. Analysis of staffing needs (essential vs. nonessential) and possible mitigations (accommodations for families and pets) go a long way to improving hospital resilience. Some recommend having a requirement to respond in a disaster written into the staff employment agreement. Others rely on the "culture" of healthcare providers to positively respond in an emergency. A candid discussion with medical and non-medical staff in advance may also improve response.

Importance of Community Partnerships, Networking/Sharing of Essential Services

None of these resilience measures would be as effective if done in isolation. All of these measures benefit from collaboration with other organizations in the healthcare sector and other community

[33] https://www.dhs.gov/critical-infrastructure-sectors
[34] https://www.ncbi.nlm.nih.gov/pubmed/25807865

members. The Joint Commission specifically looks for preparedness measures (risk management) within a community context. Healthcare coalitions are particularly valuable for community assessments and making progress toward resilience. Information sharing is key, both before and during the emergency. Identification of potential risks (hazard vulnerability analysis) in collaboration with community partners is essential. Healthcare institutions must know their role(s) in community, county, and regional disaster plans, during the planning and response. A mutual understanding of the vulnerabilities of the other critical infrastructure systems is more valuable that just looking at healthcare vulnerabilities alone.

In addition, there are often economies of scale when it comes to preparedness. Cyber- and EMP-protected microgrids may be more economically feasible when shared among several institutions on a healthcare campus. Purchasing agreements for contingencies may be complicated by the fact that numerous other users also have contracts and even (upstream) priority arrangements, such as the U.S. government. Competing contracts for diesel fuel, for example, became a critical resource that had a negative impact during Superstorm Sandy. Fuel stored in underground tanks is unavailable in a power outage, unless a separate generator is available.

Finally, if an extended power outage was to occur, hospitals could expect a surge in activity when they themselves are operating under adverse conditions and less able to provide services. A hospital's reputation is dependent on being able to provide services when its community is in need.

What Is a Microgrid? Why Be EMP-Protected?

To keep a hospital from shutting down during a prolonged power outage when backup generator fuel is unavailable, the hospital needs to generate and store enough power on-site to maintain core functions. But, it can only do that if the local power system is protected from the same causes that could lead to the failure of the commercial power grid.[35] For this reason, the DTRA asked for the development and deployment of EMP-protected microgrids for both military bases and the hospitals they use. DTRA recognizes that commercial power grids are not designed to withstand electromagnetic threats and proposes local power systems be EMP-protected and capable of operating in island mode.

A microgrid is a local energy grid with control capability, which means it can disconnect from the traditional grid and operate autonomously (i.e., "islanding").[36] Islanding is not the most challenging or difficult operational mode.

The two most difficult operational modes are: (1) the parallel operation with the grid, which has the highest value (both technically and economically) and (2) the ability to have that system be EMP-protected so, in an EMP event, the microgrid continues operating without damage or disruption by immediately detaching itself from the grid and operating in island mode indefinitely until the commercial power grid is restored. During normal operations, knowing when NOT to island is most important to overall grid stability and performance. On the other hand, a no-notice EMP event can damage the grid, its control systems, and the hospital infrastructure it serves within nanoseconds without the knowledge of grid operators. This makes it necessary for EMP-protected systems to continue to operate through and after an EMP attack in island-mode even while the larger grid ceases to function

[35] Minimizing exposure to EMP and cyberthreats from the earliest design stages can minimize the cost of protection.

[36] Smaller microgrids, sometimes known as nanogrids, can be operated off-grid without a grid connection.

after becoming one of the conductive pathways of damaging EMPs (see the DTRA press release for official comments on this threat and related IEEE[37] and military standards[38]).

Power modeling and management software, combined with network management systems, can provide the foundation for a microgrid that provides resilience from EMP and cyberthreats. It also makes it possible to monitor systems so they can recover quicker from disruptions.

The technical advisors to this handbook propose a microgrid plan that takes industry and military standards into account and recommends a microgrid system capable of meeting those standards and ultimately capable of operating either in grid and island mode.

Potential power savings/cost savings. Hospitals can fund a substantial if not all of the costs for EMP- and cyber-resilient microgrids from a combination of energy savings and energy efficiency. Additional funding might come from projects that create mutual advantages between hospitals and their strategic partners, including those who may wish to conduct further applied research in this area. The financial advantage of a microgrid is that it can operate the base power load of hospitals on a day-to-day basis while reducing the spending for that amount of power from outside utilities. Furthermore, when a utility-based power outage occurs, the hospitals can shed loads that are not essential while continuing to operate their microgrids. Emergency generators would only need to continue operation activities beyond essential loads or be held in reserve in case of any other technical difficulties. By providing their own power for essential loads (i.e., often 30%–50% of the entire power requirement), hospitals can use the savings from that portion of their power bills to fund the growth of their own systems. In many cases, though, the cost savings of local energy generation and storage may not be large enough to completely fund a microgrid. In these cases, it would be beneficial to find additional savings in energy efficiency measures so those savings could also be applied to the overall financial requirements for the on-site microgrid.

Additional reasons for on-site energy/power storage. On-site energy storage can provide cost savings that help fund the microgrid and the ability to continue functioning during a long-term power outage. It is also possible to develop a microgrid system that simultaneously serves other functions. The initial microgrid can be used to gather data to substantiate additional energy savings and performance advantages. That same data can assist in choosing the best microgrid power generation and storage elements for specific hospital requirements, environments, and utilities that it might connect. It can be used as a facility-wide power-management system that helps operators recover sooner after an

---

[37] IEEE Standard 2030.(7)/.(8) is in development at this time and will address establishing a standard for the Microgrid Energy Management System (MEMS) to enable interoperability of the different controllers and components needed to operate the MEMS through cohesive and platform-independent interfaces. The standard includes the control functions that define the microgrid as a system that can manage itself, and operate autonomously or grid connected, and seamlessly connect to and disconnect from the main distribution grid for the exchange of power and the supply of ancillary services. Additionally, the standard addresses the technical issues and challenges associated with the proper operation of the MEMS that are common to all microgrids, regardless of topology, configuration or jurisdiction, and to present the control approaches required from the distribution system operator and the microgrid operator.

[38] MIL STD 188.125 is an example of a military standard that can readily be applied to small on-site microgrids, control rooms and communication networks inherent in a microgrid. DHS has provided guidance based on that standard. (Large regional microgrids have additional complexities within its generation and transmission system that require additional work.) See: http://www.stop-emp.com/Final-DTRA-IAN-Press-Release%20-w-logos-centered.pdf

infrastructure interruption or failure. It can also be used to support EMP-protected data and facility security system elements. These goals are best served when power monitoring and modeling software assist in these tasks with the support of professional engineers.

Preliminary microgrid assessments can be accomplished to create an initial EMP-protected microgrid that can be grown over time as the initial microgrid system provides data that support the self-funding for microgrid expansion.

How to Conduct a "Lights Out" Exercise

"Lights Out" was the name of a tabletop exercise conducted to explore the consequences of a long-term widespread power outage on a local hospital. A number of subject matter experts provided background information, including speakers from: NOAA's Space Weather Prediction Center, the U.S. National Renewable Energy Labs, public safety, public health, local (county) government, and the military. The scenario used was one in which a severe space weather event was followed by an opportunistic cyberattack on the power grid. This scenario was thought to be plausible and had been vetted in the past. Short vignettes representing several days of elapsed time were followed by a series of questions for participants. The scenario, vignettes, and questions are contained in the attached documents at the end of the handbook. Speakers familiar with these subjects may be available for future exercises from InfraGard EMP-SIG, by contacting igempsig@infragardmembers.org or the Protective Security Advisor program at the U.S. Department of Homeland Security (https://www.dhs.gov/protective-security-advisors).

Appendices and Diagrams

Checklist for Hospitals in Advance of a Prolonged/Widespread Power Outage[39]

- What are the essential functions of your hospital in day-to-day operations? During an emergency? During an emergency or power outage lasting more than 30 days?
- What items from your entire supply chain need to be stocked locally? Have they been prioritized? Power is an example of what would be the most important and difficult to replace. Which others take longer to replace?
- Have you considered the use of on-site, renewable sources of electricity (solar/wind), geothermal, etc., as well as battery storage or similar energy storage system? What protection levels have they been certified from threats such as EMP and cyberattacks?
- Are priority areas/functions supplied by an EMP and cyber resilient microgrid or emergency power outlets/circuits? Partial or complete?
- Are your emergency generator(s) sufficient? How long can you continue to provide power to support essential functions without refueling?
- Are your generators sufficiently "hardened" to be able to withstand a weather emergency? A minor earthquake? Solar storm? Cyberattack? EMP? Physical damage from outside/inside your facility?
- Are the circuitry, panels, fuel tank sensors, and generators themselves, all protected from flooding damage?
- Is your plan for refueling emergency generators guaranteed and not subject to upstream requirements?
- Are you able to service a single generator (filters, oil) while keeping essential services online?
- Do you have external connections already in place for additional emergency generators?
- Are your heating (and cooling) systems on emergency power? Are there methods in place to transport patients within your facility, not electricity dependent? Is food refrigeration on emergency power?
- Are water distribution systems dependent on electricity to maintain pressure? Can you supply your water needs with a storage tank (or well) and emergency power generation? Do you have a way to purify that water if needed?
- Do your wastewater and sewage systems continue to function during a power outage? Do you have a means of preventing backflow of sewage into your facility, if required?
- Do you have several means of redundant communications, not dependent on grid power or a functioning Internet? Do other healthcare facilities in your area have similar means/capabilities?
- Can your accounting/billing, electronic medical records, and data storage continue to operate? Is there off-site backup on emergency power or onsite microgrids independent of remote resources?
- Are you able to transition to a paper system for essential record keeping functions?
- Have you arranged "come to work requirements" with essential clinical and nonclinical staff and made provision for their transportation, food and sleeping arrangements, child care, pet care, elder care, etc.

[39] For the complete HHS Sustainable and Climate Resilient Healthcare Facility Initiative Element 3 Checklist, visit https://toolkit.climate.gov/sites/default/files/SCRHCFI%20Checklist%203%2008141 5_Form.pdf

Additional Resources and References for Resilient Hospitals Handbook Users

Hospital Emergency Preparedness and Response during Super Storm Sandy: https://oig.hhs.gov/oei/reports/oei-06-13-00260.pdf

Disaster Information Management Research Center (DIMRC): http://disasterinfo.nlm.nih.gov

InfraGard National EMP Special Interest Group (sponsored by the FBI) Publications with linked bibliographies

Triple Threat Power Grid Exercise: https://www.amazon.com/Triple-Threat-Power-Grid-Exercise/dp/1633912493/

Powering Through, From Fragile Infrastructures to Community Resilience: https://www.amazon.com/Powering-Through-Infrastructures-Community-Resilience/dp/0998384402

High-Impact Threat Series of Conference Proceedings References

Planning Resilience for High-Impact Threats to Critical Infrastructure: Conference Proceedings InfraGard National EMP SIG Sessions at the 2014 Dupont Summit: https://www.amazon.com/Planning-Resilience-High-Impact-Critical-Infrastructure/dp/1633912612/

Engaging Communities for High-Impact Threats to Critical Infrastructure: Dupont Summit 2015 Conference Proceedings of the InfraGard National EMP SIG Sessions: https://www.amazon.com/Engaging-Communities-High-Impact-Critical-Infrastructure/dp/1633914291/

Mitigating High-Impact Threats to Critical Infrastructure: Conference Proceedings of the 2013 InfraGard National EMP SIG Sessions at the Dupont Summit: https://www.amazon.com/Mitigating-High-Impact-Threats-Critical-Infrastructure/dp/1633911330/

High-Impact Threats to Critical Infrastructure: Emerging Policy and Technology: https://www.amazon.com/High-Impact-Threats-Critical-Infrastructure/dp/1935907395/

Copies of the "Lights Out" After Action Report can be found on either of these websites:

www.EMPcenter.org

www.hospitalmicrogrid.com

www.hospitalCOOPhandbook.com

www.stop-EMP.com

Contributors

Charles (Chuck) Manto, President Instant Access Networks, Chair EMP Special Interest Group InfraGard (cmanto@stop-EMP.com); Earl Motzer, Ph.D., Chair, Healthcare and Public Health Sector Coordinating Council; Art Glynn, CAPT USN (Ret.), Navy Emergency Preparedness Liaison Officer, FEMA Region 8, and Consultant with Booz Allen; James Terbush, MD, MPH, Public Health Instructor and President, Innovative Health Systems Inc.

Acknowledgments

We wish to thank Catherine Feinman, Editor-in-Chief, www.DomesticPreparedness.com; and, staff of the Policy Studies Organization, for their review of this document. We also wish to thank staff at Doctors Community Hospital in Lanham, MD; the University of Kentucky Medical Center in Lexington, KY; St. Francis Hospital in Colorado Springs, CO, and the Society for Disaster Medicine and Public Health for their participation in hospital microgrid assessments and conferences on these topics that have helped inform this handbook. We are especially grateful for the offices of the state of Maryland that have also provided facilitation on these issues impacting hospitals and the National Guard that include the Maryland Emergency Management Agency, the Maryland Governor's Office of Homeland Security, and the Maryland Department of General Services, especially the active involvement of Deputy Secretary for Energy, A. Leigh Williams, Esq.

About FBI InfraGard and the EMP Special Interest Group (EMP SIG)

The Federal Bureau of Investigation's (FBI) InfraGard is a partnership between the FBI and members of the private sector to help protect the nation's infrastructure. The InfraGard program provides a vehicle for seamless public–private collaboration with government that expedites the timely exchange of information and promotes mutual learning opportunities relevant to the protection of critical infrastructure.

The EMP SIG Mission is to facilitate information sharing and subject matter expertise nationwide to serve local communities so that they may become more resilient in light of threats that would lead to catastrophic, cascading losses of life-sustaining infrastructures and resources.

Statement for the Record from the Commission to Assess the Threat to the United States from Electromagnetic Pulse (EMP) Attack

STATEMENT FOR THE RECORD
DR. WILLIAM R. GRAHAM, CHAIRMAN
DR. PETER VINCENT PRY, CHIEF OF STAFF
COMMISSION TO ASSESS THE THREAT TO THE UNITED STATES FROM ELECTROMAGNETIC PULSE (EMP) ATTACK
TO
U.S. HOUSE OF REPRESENTATIVES
COMMITTEE ON HOMELAND SECURITY
SUBCOMMITTEE ON OVERSIGHT AND MANAGEMENT EFFICIENCY
HEARING
"EMPTY THREAT OR SERIOUS DANGER: ASSESSING NORTH KOREA'S RISK TO THE HOMELAND"
October 12, 2017

North Korea Nuclear EMP Attack: An Existential Threat

During the Cold War, major efforts were undertaken by the Department of Defense to assure that the U.S. national command authority and U.S. strategic forces could survive and operate after an EMP attack. However, no major efforts were then thought necessary to protect critical national infrastructures, relying on nuclear deterrence to protect them. With the development of small nuclear arsenals and long-range missiles by new, radical U.S. adversaries, beginning with North Korea, the threat of a nuclear EMP attack against the U.S. becomes one of the few ways that such a country could inflict devastating damage to the United States. It is critical, therefore, that the U.S. national leadership address the EMP threat as a critical and existential issue, and give a high priority to assuring the leadership is engaged and the necessary steps are taken to protect the country from EMP.

By way of background, the Commission to Assess the Threat to the United States from Electromagnetic Pulse (EMP) Attack was established by Congress in 2001 to advise the Congress, the President, Department of Defense and other departments and agencies of the U.S. Government on the nuclear EMP threat to military systems and civilian critical infrastructures. The EMP Commission was re-established in 2015 with its charter broadened to include natural EMP from solar storms, all manmade EMP threats, cyber-attack, sabotage and Combined-Arms Cyber Warfare. The EMP Commission charter gives it access to all relevant classified and unclassified data and the power to levy analysis upon the Department of Defense.

On September 30, 2017, the Department of Defense, after withholding a significant part of the monies allocated by Congress to support the work of the EMP Commission for the entirety of 2016, terminated funding the EMP Commission. In the same month, North Korea detonated an H-Bomb that it plausibly describes as capable of "super-powerful EMP" attack and released a technical report "The EMP Might of Nuclear Weapons" accurately describing what Russia and China call a "Super-EMP" weapon.

Neither the Department of Defense nor the Department of Homeland Security has asked Congress to continue the EMP Commission. The House version of the National Defense Authorization Act includes a provision that would replace the existing EMP Commission with new Commissioners. Yet the existing EMP Commission comprises the nation's foremost experts who have been officially or unofficially continuously engaged trying to advance national EMP preparedness for 17 years.

And today, as the EMP Commission has long warned, the nation faces a potentially imminent and existential threat of nuclear EMP attack from North Korea. Recent events have proven the EMP Commission's critics wrong about other highly important aspects of the nuclear missile threat from North Korea:

--Just six months ago, most experts thought North Korea's nuclear arsenal was primitive, some academics claiming it had as few as 6 A-Bombs. Now the intelligence community reportedly estimates North Korea has 60 nuclear weapons.
--Just six months ago, most experts thought North Korea's ICBMs were fake, or if real could not strike the U.S. mainland. Now the intelligence community reportedly estimates North Korea's ICBMs can strike Denver and Chicago, and perhaps the entire United States.
--Just six months ago, most experts thought North Korea was many years away from an H-Bomb. Now it appears North Korea has H-Bombs comparable to sophisticated U.S. two-stage thermonuclear weapons.
--Just six months ago, most experts claimed North Korean ICBMs could not miniaturize an A-Bomb or design a reentry vehicle for missile delivery. Now the intelligence community reportedly assesses North Korea has miniaturized nuclear weapons, and has developed reentry vehicles for missile delivery, including by ICBMs that can strike the U.S.[1]

After massive intelligence failures grossly underestimating North Korea's long-range missile capabilities, number of nuclear weapons, warhead miniaturization, and proximity to an H-Bomb, the biggest North Korean threat to the U.S. remains unacknowledged—nuclear EMP attack.

North Korea confirmed the EMP Commission's assessment by testing an H-Bomb that could make a devastating EMP attack, and in its official public statement: "The H-Bomb, the explosive power of which is adjustable from tens of kilotons to hundreds of kilotons, is a multi-functional thermonuclear weapon with great destructive power which can be detonated even at high altitudes for super-powerful EMP attack according to strategic goals."[2]

[1] Joby Warwick, Ellen Nakashima, Anna Fifield, "North Korea Is No Making Missile-Ready Nuclear Weapons, U.S. Analysts Say" Washington Post, August 18, 2017; Michelle Ye Hee Lee, "North Korean Nuclear Test May Have Been Twice As Strong As First Thought" Washington Post, September 13, 2017; Jack Kim, Soyoung Kim, "North Korea Says It Has Developed A More Advanced Hydrogen Bomb That Can Be Loaded Onto An ICBM" Business Insider, September 2, 2017; NBC News, "'A Big Hoax': Experts Say North Korea Showing Off Missiles That Can't Fly" August 15, 2013.

[2] Bill Gertz, "Korea Nuclear Test Furthers EMP Bomb" Washington Free Beacon, September 6, 2017.

As noted earlier, Pyongyang also released a technical report accurately describing a "Super-EMP" weapon.[3]

Just six months ago, some academics dismissed EMP Commission warnings and even, literally, laughed on National Public Radio at the idea North Korea could make an EMP attack.

Primitive and "Super-EMP" Nuclear Weapons are Both EMP Threats

The EMP Commission finds that even primitive, low-yield nuclear weapons are such a significant EMP threat that rogue states, like North Korea, or terrorists may well prefer using a nuclear weapon for EMP attack, instead of destroying a city: "Therefore, terrorists or state actors that possess relatively unsophisticated missiles armed with nuclear weapons may well calculate that, instead of destroying a city or military base, they may obtain the greatest political-military utility from one or a few such weapons by using them—or threatening their use—in an EMP attack."[4]

The EMP Commission 2004 Report warns: "Certain types of relatively low-yield nuclear weapons can be employed to generate potentially catastrophic EMP effects over wide geographic areas, and designs for variants of such weapons may have been illicitly trafficked for a quarter-century."[5]

In 2004, two Russian generals, both EMP experts, warned the EMP Commission that the design for Russia's Super-EMP warhead, capable of generating high intensity EMP fields over 100,000 volts per meter, was "accidentally" transferred to North Korea. They also said that due to "brain drain," Russian scientists were in North Korea, as were Chinese and Pakistani scientists according to the Russians, helping with the North's missile and nuclear weapon programs. In 2009, South Korean military intelligence told their press that Russian scientists are in North Korea helping develop an EMP nuclear weapon. In 2013, a Chinese military commentator stated North Korea has Super-EMP nuclear weapons.[6]

Super-EMP weapons are low-yield and designed to produce not a big kinetic explosion, but rather a high level of gamma rays, which generates the high-frequency E1 EMP that is most damaging to the broadest range of electronics. North Korean nuclear tests, including the first in 2006, whose occurrence was predicted to the EMP Commission two years in advance by the two

---

[3] Ibid. Kim Song-won, Dean of Kim Chaek University of Technology "The EMP Might of Nuclear Weapons" Rodong Sinmun, Pyongyang, September 4, 2017.

[4] Commission to Assess the Threat to the United States from Electromagnetic Pulse (EMP) Attack, *Executive Report*, 2004, p. 2.

[5] Ibid.

[6] U.S. Senate, Hearing, Statement for the Record, Dr. Peter Vincent Pry, "Foreign Views of Electromagnetic Pulse (EMP) Attack" testimony on behalf of EMP Commission before the Subcommittee on Terrorism, Technology, and Homeland Security, Senate Committee on the Judiciary (Washington, D.C.: March 9, 2005); Kim Min-sek and Yoo Jee-ho, "Military Source Warns of North's EMP Bomb" JoonAng Daily (September 2, 2009); Li Daguang, "North Korean Electromagnetic Attack Threatens South Korea's Information Warfare Capabilities" Tzu Chin, No. 260 (June 1, 2012) pp. 44-45.

Russian EMP experts, mostly have yields consistent with the size of a Super-EMP weapon. The Russian generals' accurate prediction about when North Korea would perform its first nuclear test, and of a yield consistent with a Super-EMP weapon, indicates their warning about a North Korean Super-EMP weapon should be taken very seriously.

EMP Threat From Satellites

While most analysts are fixated on when in the future North Korea will develop highly reliable intercontinental missiles, guidance systems, and reentry vehicles capable of striking a U.S. city, the threat here and now from EMP is largely ignored. EMP attack does not require an accurate guidance system because the area of effect, having a radius of hundreds or thousands of kilometers, is so large. No reentry vehicle is needed because the warhead is detonated at high-altitude, above the atmosphere. Missile reliability matters little because only one missile has to work to make an EMP attack against an entire nation.

North Korea could make an EMP attack against the United States by launching a short-range missile off a freighter or submarine or by lofting a warhead to 30 kilometers burst height by balloon. While such lower-altitude EMP attacks would not cover the whole U.S. mainland, as would an attack at higher-altitude (300 kilometers), even a balloon-lofted warhead detonated at 30 kilometers altitude could blackout the Eastern Electric Power Grid that supports most of the population and generates 75 percent of U.S. electricity.

Or an EMP attack might be made by a North Korean satellite, right now.

A Super-EMP weapon could be relatively small and lightweight, and could fit inside North Korea's Kwangmyongsong-3 (KMS-3) and Kwangmyongsong-4 (KMS-4) satellites. These two satellites presently orbit over the United States, and over every other nation on Earth--demonstrating, or posing, a potential EMP threat against the entire world.

North Korea's KMS-3 and KMS-4 satellites were launched to the south on polar trajectories and passed over the United States on their first orbit. Pyongyang launched KMS-4 on February 7, 2017, shortly after its fourth illegal nuclear test on January 6, that began the present protracted nuclear crisis with North Korea.

The south polar trajectory of KMS-3 and KMS-4 evades U.S. Ballistic Missile Early Warning Radars and National Missile Defenses, resembling a Russian secret weapon developed during the Cold War, called the Fractional Orbital Bombardment System (FOBS) that would have used a nuclear-armed satellite to make a surprise EMP attack on the United States.[7]

Ambassador Henry Cooper, former Director of the U.S. Strategic Defense Initiative, and a preeminent expert on missile defenses and space weapons, has written numerous articles warning

---

[7] Miroslav Gyurosi, *The Soviet Fractional Orbital Bombardment System Program*, (January 2010) Technical Report APA-TR-2010-010.

about the potential North Korean EMP threat from their satellites. For example, on September 20, 2016 Ambassador Cooper wrote:

U.S. ballistic missile defense (BMD) interceptors are designed to intercept a few North Korean ICBMs that approach the United States over the North Polar region. But current U.S. BMD systems are not arranged to defend against even a single ICBM that approaches the United States from over the South Polar region, which is the direction toward which North Korea launches its satellites...This is not a new idea. The Soviets pioneered and tested just such a specific capability decades ago—we call it a Fractional Orbital Bombardment System (FOBS)...So, North Korea doesn't need an ICBM to create this existential threat. It could use its demonstrated satellite launcher to carry a nuclear weapon over the South Polar region and detonate it...over the United States to create a high-altitude electromagnetic pulse (HEMP)...The result could be to shut down the U.S. electric power grid for an indefinite period, leading to the death within a year of up to 90 percent of all Americans—as the EMP Commission testified over eight years ago.[8]

Former NASA rocket scientist James Oberg visited North Korea's Sohae space launch base, witnessed elaborate measures undertaken to conceal space launch payloads, and concludes in a 2017 article that the EMP threat from North Korea's satellites should be taken seriously:

...there have been fears expressed that North Korea might use a satellite to carry a small nuclear warhead into orbit and then detonate it over the United States for an EMP strike. These concerns seem extreme and require an astronomical scale of irrationality on the part of the regime. The most frightening aspect, I've come to realize, is that exactly such a scale of insanity is now evident in the rest of their 'space program." That doomsday scenario, it now seems, has been plausible enough to compel the United States to take active measures to insure that no North Korean satellite, unless thoroughly inspected before launch, be allowed to reach orbit and ever overfly the United States.[9]

Kim Jong-Un has threatened to reduce the United States to "ashes" with "nuclear thunderbolts" and threatened to retaliate for U.S. diplomatic and military pressure by "ordering officials and scientists to complete preparations for a satellite launch as soon as possible" amid "the enemies'

---

[8] Ambassador Henry F. Cooper, "Whistling Past The Graveyard..." High Frontier (September 20, 2016) highfrontier.org/sept-20-2016-whistling-past-the-graveyard/ See also: highfrontier.org/category/fobs. On up to 90% U.S. fatalities from an EMP attack, during a congressional hearing, Rep. Roscoe Bartlett asked me if such high fatalities could result, and I responded: "We don't have experience with losing the infrastructure in a country with 300 million people, most of whom don't live in a way that provides for their own food and other needs. We can go back to an era when people did live like that. That would be—10 percent would be 30 million people, and that is probably the range where we could survive as a basically rural economy." U.S. House of Representatives, Hearing, "Threat Posed By Electromagnetic Pulse (EMP) Attack" Committee on Armed Services (Washington, D.C.: July 10, 2008), p. 9.

[9] Jim Oberg, Space Review (February 6, 2017) www.thespacereview.com/article/3164/1in a 2017 aricle

harsh sanctions and moves to stifle" the North.[10] North Korean press (for example in Rodong Sinmun; March 7, 2016) asserts readiness for "any form of war" and includes their satellite with "strengthening of the nuclear deterrent and legitimate artificial satellite launch, which are our fair and square self-defensive choice." Moreover: "The nuclear [weapons] we possess are, precisely, the country's sovereignty, right to live, and dignity. Our satellite that cleaves through space is the proud sign that unfolds the future of the most powerful state in the world." The same article, like many others, warns North Korea makes "constant preparations so that we can fire the nuclear warheads, which have been deployed for actual warfare for the sake of national defense, at any moment!"

An earlier generation immediately understood the alarming strategic significance of Sputnik in 1957, yet few today understand or even care about the strategic significance of North Korea's satellites, perhaps because of widespread ignorance about EMP.

Addressing Misinformation

Misinformation about EMP abounds in the media, and even in many allegedly serious studies, from uninformed persons posturing as experts, who have no competency in EMP. False claims are often made that the EMP threat is "not real" but merely theoretical and greatly overblown.[11]

For example, one academic often quoted by the press claims that during the 1962 STARFISH PRIME high-altitude nuclear test, "just one string of street lights failed in Honolulu" and that this proved EMP is no threat.[12] In fact, the EMP knocked-out 36 strings of street lights, caused a telecommunications microwave relay station to fail, burned out HF (High-Frequency) radio links (used for long-distance communications), set off burglar alarms, and caused other damage.[13]

The Hawaiian Islands did not experience a catastrophic protracted blackout because they were on the far edge of the EMP field contour, where effects are weakest; are surrounded by an ocean, which mitigates EMP effects; and were still in an age dominated by vacuum tube electronics.

STARFISH PRIME was not the only test of this kind. Russia in 1961-62 also conducted a series of high-altitude nuclear bursts to test EMP effects over Kazakhstan, an industrialized area nearly

---

[10] Alex Lockie, "North Korea Threatens 'Nuclear Thunderbolts' As U.S. And China Finally Work Together" American Military News (April 14, 2017); Fox News, "U.S. General: North Korea 'Will' Develop Nuclear Capabilities To Hit America" (September 20, 2016) www.foxnews.com/world/2016/09/20/north-korea-says-successfully-ground-tests-new-rocket-engine.html

[11] See for example: Jeffrey Lewis, "Would A North Korean Space Nuke Really Lay Waste to the U.S.?" New Scientist www.newscientist.com/article/2129618; Lewis quoted in Cheyenne MacDonald, "A North Korean 'Space Nuke' Wouldn't Lay Waste To America" Daily Mail, May 3, 2017; Lewis interviewed by National Public Radio, "The North Korean Electromagnetic Pulse Threat, Or Lack Thereof" www.npr.org/2017/04/27/525833275; www.naturalnews.com/2017-05-01-npr-laughs-hysterically-north-korean-emp-nuclear-attack.html

[12] Ibid.

[13] Dr. William R. Graham, "North Korean Nuclear EMP Attack: An Existential Threat" 38 North, June 2, 2017.

as large as Western Europe.[14] That test destroyed the Kazakh electric grid.[15] Moreover, modern electronics, in part because they are designed to operate at much lower voltages, are much more vulnerable to EMP than the electronics of 1962 exposed to STARFISH PRIME and the Kazakh nuclear tests. A similar EMP event over the U.S. today would be an existential threat.[16]

Another academic wrongly asserts that because EMP from atmospheric nuclear tests in Nevada did not blackout Las Vegas, therefore EMP is no threat. The nuclear tests he describes were all endo-atmospheric tests that do not generate appreciable EMP fields beyond a range of about 5 miles. The high-altitude EMP (HEMP) threat of interest requires exo-atmospheric detonation, at 30 kilometers altitude or above, and produces EMP out to ranges of hundreds to thousands of miles. Las Vegas was not affected by the Nevada tests because they were endo-atmospheric nuclear tests that generated no HEMP.[17]

The same academic also miscalculates that "a 20-kiloton bomb detonated at optimum height would have a maximum EMP damage distance of 20 kilometers" in part, because he assumes "15,000 volts/meter or higher" in the E1 EMP component is necessary for damage. This figure is an extreme overestimate of system damage field thresholds. Damage and upset to electronic systems will happen from E1 EMP field strengths far below the academic's "15,000 volts/meter or higher." A one meter wire connected to a semiconductor device, such as a mouse cord or interconnection cable, would place hundreds to thousands of volts on microelectronic devices out to ranges of hundreds of miles for low-yield nuclear devices. Based on omission and other experience with many EMP tests, semiconductor junctions, operating at a few volts, will experience breakdown at a few volts over their operating point, allowing their power supply to destroy the junctions experiencing breakdown.[18]

The same academic and many other non-experts also ignore system upset as a vulnerability. Digital electronics can be upset by extraneous pulses of a few volts. For unmanned control systems present within the electric power grids, long-haul communication repeater stations, and gas pipelines, an electronic upset is tantamount to permanent damage. Temporary upset of electronics can also have catastrophic consequences for military operations. No electronics should be considered invulnerable to EMP unless hardened and tested to certify survivability.

---

[14] High-altitude EMP (HEMP), the phenomenon under discussion, results from the detonation of a nuclear weapon at high-altitude, 30 kilometers or higher. All nuclear weapons, even a primitive Hiroshima-type A-bomb, can produce levels of HEMP damaging to modern electronics over large geographic regions.

[15] According to Electric Infrastructure Security Council, *Report: USSR Nuclear EMP Upper Atmosphere Kazakhstan Test 184*, (www.eiscouncil.org/APP_Data/upload/a4ce4b06-1a77-44d-83eb-842bb2a56fc6.pdf), citing research by Oak Ridge National Laboratory, a comparable EMP event over the U.S. today "would likely damage about 365 large transformers in the U.S. power grid, leaving about 40 percent of the U.S. population without electrical power for 4 to 10 years."

[16] EMP Commission *Executive Report*, op. cit., pp. 4-8.

[17] Jack Liu, "A North Korean EMP Attack?...Unlikely" 38 North, May 5, 2017.

[18] Ibid.

Some highly critical unprotected electronics have been upset or damaged in simulated EMP tests, not at "15,000 volts/meter or higher," but at threat levels far below 1,000 volts/meter.[19]

The North Korean missile test on April 29, 2017, which apparently detonated at an altitude of 72 kilometers, the optimum height-of-burst for EMP attack by a 10 KT warhead, would create a potentially damaging EMP field spanning, not the academic's miscalculated 20 kilometers radius, but to about 930 kilometers radius [Kilometers Radius = 110 (Kilometers Burst Height to the 0.5 Power)].[20]

Therefore, even for a low-yield 10-20 kiloton weapon, the EMP field should be considered dangerous for unprotected U.S. systems. The EMP Commission 2004 Report warned against the U.S. military's increasing use of commercial-off-the-shelf-technology that is not protected against EMP: "Our increasing dependence on advanced electronics systems results in the potential for an increased EMP vulnerability of our technologically advanced forces, and if unaddressed makes EMP employment by an adversary an attractive asymmetric option."[21]

Empirical Basis for EMP Threat Better Established than Cyber Threat

The empirical basis for the threat of an EMP attack to electric grids and other critical infrastructures is far deeper and broader than the data for cyber-attacks or sabotage. The notion that a cyber-attack or sabotage can plunge the U.S. into a protracted blackout--while very real threats that warrant deep concern--are far more theoretical constructs than EMP attack.

We know for certain that EMP will cause widespread damage of electronics and protracted blackout of unprotected electric grids and other critical infrastructures from such hard data as:

--The U.S. STARFISH PRIME high-altitude nuclear test in 1962 over Johnston Island that generated an EMP field over the Hawaiian Islands, over 1,300 kilometers away, causing widespread damage to electronic systems.[22]
--Six Russian EMP tests 1961-1962 over Kazakhstan that with a single weapon destroyed electric grids over an area larger than Western Europe, proving this capability six times.[23]
--30 years (1962-1992) of U.S. underground nuclear testing that included collecting data on EMP effects.
--Over 50 years of testing by EMP simulators, still ongoing, including by the Congressional EMP Commission (2001-2008) that proved modern electronics are over 1 million times more vulnerable to EMP than the electronics of 1962.[24]

---

[19] Ibid.
[20] Ibid.
[21] EMP Commission, *Executive Report*, op. cit., p. 47.
[22] Phil Plait, "The 50th Anniversary of Starfish Prime: The Nuke That Shook The World" Discover, July 9, 2012.
[23] Jerry Emanuelson, "Soviet Test 184: The 1962 Soviet Nuclear EMP Tests Over Kazakhstan" Future Science, Undated; Vladimir M. Loborev, "Up to Date State of the NEMP Problems and Topical Research Directions" Electromagnetic Environments and Consequences: Proceedings of the European International Symposium on Electromagnetic Environments, EUROEM Conference, Bordeaux, France, 1994; V. N. Mikhailov, *The Nuclear Tests of the USSR*, Vol. 2, Institute of Strategic Stability, Rosatom.

Moreover, hard data proving the threat from nuclear EMP is available from natural EMP generated by geomagnetic storms, accidental damage caused by electromagnetic transients, and non-nuclear radiofrequency weapons (RF weapons). All of these produce field strengths much less powerful than nuclear EMP, and in the case of accidental electromagnetic transients and radiofrequency weapons, much more localized. There are many thousands of such cases.

Many documented examples of successful attacks using RF weapons, and accidents involving electromagnetic transients, are described in the Department of Defense *Pocket Guide for Security Procedures and Protocols for Mitigating Radio Frequency Threats* (Technical Support Working Group, Directed Energy Technical Office, Dahlgren Naval Surface Warfare Center). A few examples:

--"Radio Frequency Weapons were used in separate incidents against the U.S. Embassy in Moscow to falsely set off alarms and to induce a fire in a sensitive area."
--"In Kzlyar, Dagestan, Russia, Chechen rebel commander Salman Raduyev disabled police radio communications using RF transmitters during a raid."
--"In June 1999 in Bellingham, Washington, RF energy from a radar induced a SCADA malfunction that caused a gas pipeline to rupture and explode."
--"In 1999, a Robinson R-44 news helicopter nearly crashed when it flew by a high-frequency broadcast antenna."
--North Korea used a Radio Frequency Weapon, purchased from Russia, to attack airliners and impose an "electromagnetic blockade" on air traffic to Seoul, South Korea's capital. The repeated attacks by RFW also disrupted communications and the operation of automobiles in several South Korean cities in December 2010; March 9, 2011; and April-May 2012.[25]

Vulnerabilities to EMP

When assessing the potential vulnerability of U.S. military forces and civilian critical infrastructures to EMP, it is necessary to be mindful of the complex interdependencies of these highly-networked systems, because EMP upset and damage of a very small fraction of the total system can cause total system failure.[26]

Real world failures of electric grids from various causes indicate that a nuclear EMP attack would have catastrophic consequences. Significant and highly disruptive blackouts have been caused by single-point failures cascading into system-wide failures, originating from damage comprising far less than 1 percent of the total system. For example:

--The Great Northeast Blackout of 2003--that put 50 million people in the dark for a day, contributed to at least 11 deaths, and cost an estimated $6 billion—originated from a single

---

[24] "Electromagnetic Pulse: Threat to Critical Infrastructures" Hearing before the Subcommittee on Cybersecurity, Infrastructure Protection, and Security Technologies, House Committee on Homeland Security, Washington, D.C.: May 8, 2014.

[25] "Massive GPS Jamming Attack By North Korea" GPSWORLD.COM, May 8, 2012.

[26] Report of the Commission to Assess the Threat to the United States from Electromagnetic Pulse (EMP) Attack, *Critical National Infrastructures*, 2008, passim.

failure point when a powerline contacted a tree branch, damaging less than 0.0000001 (0.00001%) of the system.
--The New York City Blackout of 1977, that resulted in the arrest of 4,500 looters and injury of 550 police officers, was caused by a lightning strike on a substation that tripped two circuit breakers.
--The Great Northeast Blackout of 1965, that effected 30 million people, happened because a protective relay on a transmission line was improperly set.
--India's nationwide blackout of July 30-31, 2012—the largest blackout in history, effecting 670 million people, 9% of the world population—was caused by overload of a single high-voltage powerline.
--India's blackout of January 2, 2001—effecting 226 million people—was caused by equipment failure at the Uttar Pradesh substation.
--Indonesia's blackout of August 18, 2005—effecting 100 million people—was caused by overload of a high-voltage powerline.
--Brazil's blackout of March 11, 1999—effecting 97 million people—was caused by a lightning strike on an EHV transformer substation.
--Italy's blackout of September 28, 2003—effecting 55 million people—was caused by overload of two high-voltage powerlines.
--Germany, France, Italy, and Spain experienced partial blackouts on November 4, 2006—effecting 10-15 million people—from accidental shutdown of a high-voltage powerline.
--The San Francisco blackout in April 2017 was caused by the failure of a single high voltage breaker.

In contrast to the above blackouts caused by single-point or small-scale failures, a nuclear EMP attack would inflict massive widespread damage to the electric grid causing millions of failure points. With few exceptions, the U.S. national electric grid is unhardened and untested against nuclear EMP attack.

In the event of a nuclear EMP attack on the United States, a widespread protracted blackout is inevitable. This commonsense assessment is also supported by the nation's best computer modeling:

--Modeling by the U.S. Federal Energy Regulatory Commission (FERC) reportedly assesses that a terrorist attack that destroys just 9 of 2,000 EHV transformers--merely 0.0045 (0.45%) of all EHV transformers in the U.S. national electric grid--would be catastrophic damage, causing a protracted nationwide blackout.
--Modeling by the Congressional EMP Commission assesses that a terrorist nuclear EMP attack, using a primitive 10-kiloton nuclear weapon, could destroy dozens of EHV transformers, thousands of SCADAS and electronic systems, causing catastrophic collapse and protracted blackout of the U.S. Eastern Grid, putting at risk the lives of millions.[27]

---

[27]For the best unclassified modeling assessment of likely damage to the U.S. national electric grid from nuclear EMP attack see: U.S. Federal Energy Regulatory Commission (FERC) Interagency Report, coordinated with the Department of Defense and Oak Ridge National Laboratory: *Electromagnetic Pulse: Effects on the U.S. Power Grid,*

Thus, even if North Korea has only primitive, low-yield nuclear weapons, and likewise if other states or terrorists acquire one or a few such weapons, and the capability to detonate them at 30 kilometers or higher-altitude over the United States, as the EMP Commission warned over a decade ago in its 2004 Report: "The damage level could be sufficient to be catastrophic to the Nation, and our current vulnerability invites attack."[28]

What Is To Be Done?

We recommend establishing an Executive Agent – a Cabinet Secretary designated by the President – with the authority, accountability, and resources, to manage U.S. national infrastructure protection and defense against EMP and the other existential threats described above. Current institutional authorities and responsibilities–government, industry, regulatory agencies—are fragmented, incomplete, and unable to protect and defend against foreign hostile EMP threats or solar super-storms.

We encourage the President to work with Congressional leaders to stand-up an *ad hoc* Joint Presidential-Congressional Commission, with its members charged with supporting the Nation's leadership and providing expertise, experience, and oversight to achieve, on an accelerated basis, the protection of critical national infrastructures. The U.S. Federal Energy Regulatory Commission (FERC) and North American Electric Reliability Corporation (NERC) have for nearly a decade been unable or unwilling to implement the EMP Commission's recommendations. A Presidential-Congressional Commission on Critical Infrastructure Protection could engage the Free World's preeminent experts on EMP and Combined-Arms Cyber Warfare to serve the entire Government in a manner akin to the Atomic Energy Commission of the 1947-74 period, advising the Administration's actions to attain most quickly and most cost-effectively the protection essential to long-term national survival and wellbeing. The United States should not remain in our current state of fatal vulnerability to well-known natural and man-made threats.

We highly commend President Trump's new Executive Order "Strengthening the Cybersecurity of Federal Networks and Critical Infrastructure" signed on May 11, 2017. We strongly recommend that implementation of cybersecurity for the electric grid and other critical infrastructures include EMP protection, since all-out cyber warfare as planned by Russia, China, North Korea, and Iran includes nuclear EMP attack. However, current institutional arrangements for protecting and improving the reliability of the electric grids and other critical infrastructures through the U.S. FERC and the NERC are not designed to

---

Executive Summary (2010); FERC Interagency Report by Edward Savage, James Gilbert and William Radasky, *The Early-Time (E1) High-Altitude Electromagnetic Pulse (HEMP) and Its Impact on the U.S. Power Grid* (Meta-R-320) Metatech Corporation (January 2010); FERC Interagency Report by James Gilbert, John Kappenman, William Radasky, and Edward Savage, *The Late-Time (E3) High-Altitude Electromagnetic Pulse (HEMP) and Its Impact on the U.S. Power Grid* (Meta-R-321) Metatech Corporation (January 2010).

[28] EMP Commission *Executive Report*, op. cit., p. 1.

address major national security threats to the electric power grids and other national critical infrastructures. Using FERC and NERC to achieve this level of national security is beyond the purpose for which those organizations were created and has proven to be fundamentally unworkable. New institutional arrangements are needed to advance preparedness to survive EMP and related threats to our critical national infrastructures.

We recommend that U.S. military forces and critical national infrastructures be protected from EMP as outlined in the EMP Commission's classified reports and unclassified reports provided in 2004 and 2008. EMP protection of military systems and civilian/military critical national infrastructures can be achieved cost-effectively by a combination of operational procedures and physical hardening. It is not necessary to harden everything. Selective hardening of key critical nodes and equipment will suffice. Threat parameters are 200 kilovolts/meter for E1 EMP and 85 volts/kilometer for E3 EMP. Critical national infrastructures are already adequately protected from E2 EMP, equivalent to lightning.

We recommend, given the proximity and enormity of the threat from EMP and Combined-Arms Cyber Warfare, the President exercise leadership to implement immediate, mid-term, and long-term steps to deter and defeat this existential threat:

Immediately:

We recommend that the President declare that EMP or cyber-attacks that blackout or threaten to blackout the national electric grid constitute the use of weapons of mass destruction that justify preemptive and retaliatory responses by the United States using all possible means, including nuclear weapons. Some potential adversaries have the capability to produce a protracted nationwide blackout induced by EMP or Combined-Arms Cyber Warfare by the use of nuclear or non-nuclear means. A Defense Science Board study *Resilient Military Systems and the Advanced Cyber Threat* (January 2013) equates an all-out cyber-attack on the United States with the consequences of a nuclear attack, and concludes that a nuclear response is justified to deter or retaliate for cyber warfare that threatens the life of the nation: "While the manifestation of a nuclear and cyber-attack are very different, in the end, the existential impact to the United States is the same."

We recommend that the President issue an Executive Order, provided to the previous White House, titled "Protecting the United States from Electromagnetic Pulse (EMP)". Among many other provisions to protect the nation from EMP on an emergency basis, the Executive Order would instantly mobilize a much needed "whole of government solution" to the EMP and combined-arms cyber threat: "All U.S. Government Departments, Agencies, Offices, Councils, Boards, Commissions and other U.S. Government entities...shall take full and complete account of the EMP threat in forming policies and plans to protect United States critical infrastructures..." Protecting the electric grids and other critical infrastructures from the worst threat—nuclear EMP attack—can, if carried out in a system-wide, integrated approach, help mitigate all lesser threats, including natural EMP, man-made non-nuclear EMP, cyber-attack, physical sabotage, and severe terrestrial weather.

We recommend that the President direct the Secretary of Defense to include a Limited Nuclear Option for EMP attack among the U.S. nuclear strike plans, and immediately make targeting and fusing adjustments to some of the nuclear forces needed to implement a nuclear EMP attack capability.

We recommend that the President direct the Secretary of Defense to use national technical means to ascertain if there is a nuclear weapon aboard North Korea's KMS-3 or KMS-4 satellites that orbit over the United States. If either or both of these satellites are nuclear-armed, they should be intercepted and destroyed over a broad ocean area where an EMP resulting from salvage-fusing will do the least damage to humanity.

We recommend that the President direct the Secretary of Defense to post Aegis ships in the Gulf of Mexico and near the east and west coasts, to search for and be prepared to intercept missiles launched from freighters, submarines, or other platforms that might make a nuclear EMP attack on the United States. U.S. National Missile Defenses (NMD) are primarily located in Alaska and California and oriented for a missile attack coming at the U.S. from the north, and are not deployed to intercept a short-warning missile attack launched near the U.S. coasts.

We recommend that the President direct the Secretary of Homeland Security to harden the FirstNet emergency communications system against EMP.

We recommend that the President initiate training, evaluating, and "Red Teaming" efforts to protect the U.S. and in the event of an EMP attack to respond, and periodically report the results of these efforts to the Congress.

Mid-Term:

We recommend that the President direct the Secretary of Defense to deploy Aegis-ashore missile interceptors along the Gulf of Mexico coast to plug the hole in U.S. missile defenses. The U.S. has no Ballistic Missile Early Warning System radars or missile interceptors facing south, and is largely blind and defenseless from that direction, including to missiles launched from submarines or off ships, or from a nuclear-armed satellite orbiting on a south polar trajectory.

We recommend that the President direct the Secretary of Defense to develop a space-surveillance program to detect if any satellites orbited over the United States are nuclear-armed, and develop space-interception capabilities to defend against nuclear-armed satellites that might make an EMP attack.

We recommend that the President direct the Nuclear Regulatory Commission to launch a crash program to harden the over 100 nuclear power reactors and their spent fuel storage facilities against nuclear EMP attack. Nuclear power reactors typically only have enough emergency power to cool reactor cores and spent fuel rods for a few days, after which they would "go Fukushima" spreading radioactivity over much of the United States.

Long-Term:

We recommend that the President through his Executive Agent protect elements of the national electric grids, the keystone critical infrastructure upon which all other critical infrastructures depend. Priority should be given to elements that are difficult and time-consuming to replace. Such elements can be protected from EMP at very low cost relative to the costs of an EMP catastrophe, and paid for without federal dollars by a slight increase in user electric rates. **We recommend that a similar approach be taken to key elements of the national telecommunications infrastructure and other national critical infrastructures.**

We recommend the development and deployment of enhanced-EMP nuclear weapons and other means to deter adversary attack on the United States. Enhanced-EMP nuclear weapons, called by the Russians Super-EMP weapons, can be developed without nuclear testing.

We recommend strengthening U.S. ballistic missile defenses—including deployment of space-based defenses considered by the Strategic Defense Initiative— and that these be designed and postured to also protect the U.S. from EMP attack.

Made in United States
Cleveland, OH
05 November 2026

47667166R00147